Advances in Karst Science

This book series explores advancements in the field of karst from a variety of perspectives, aiming to facilitate knowledge exchange and promote collaboration, thereby creating foundational steps in this process. It covers methodologies, monitoring techniques, data analysis, and interpretation across the diverse climatic, geological, and hydrogeological contexts in which karst occurs. Case studies are included to provide concrete examples of scientific advancements.

Key issues to be addressed include water supply, contamination, and land use management. While these issues may be localized, they share many common features on a global scale. This series serves as a critical resource for the scientific community, enabling comparisons of problems, results, and solutions. The information presented will be valuable for decision-makers involved in development planning in karst regions. Furthermore, the contributions can be used in educational settings and to collaborate with stakeholders, scientists, and engineers to determine practical applications of these advancements in addressing common global challenges.

The series aims to build a diverse library of reference works, textbooks, proceedings, and monographs that describe the current understanding of selected themes. The books in the series are authored by leading experts actively engaged in the relevant field. *Advances in Karst Science* includes both single- and multi-authored books, as well as edited volumes.

Ali Yamaç · Ezgi Tok

Underground Shelters of Cappadocia, Turkey

Springer

Ali Yamaç
OBRUK Cave Research Group
Istanbul, Türkiye

Ezgi Tok
Eurasia Earth Sciences Institute
Istanbul Technical University
Istanbul, Türkiye

ISSN 2511-2066 ISSN 2511-2082 (electronic)
Advances in Karst Science
ISBN 978-3-032-16094-2 ISBN 978-3-032-16095-9 (eBook)
https://doi.org/10.1007/978-3-032-16095-9

This Springer imprint is published by the registered company Springer Nature Switzerland AG
The registered company address is: Gewerbestrasse 11, 6330 Cham, Switzerland

Acknowledgments

We are deeply grateful to our dear friend Roberto Bixio. He always supported us with his incredible knowledge and helpfulness not only about the Cappadocia underground shelters but also about artificial cavities in almost every corner of Turkey.

This book was primarily created as a result of the "Kayseri Underground Structures Inventory Project." We express our deepest gratitude to Prof. Osman Özsoy, whose efforts transformed this project into a reality and facilitated its attainment of global prominence over the years. This project and this book would be impossible without his constant dedication and exceptional efforts.

The OBRUK Cave Research Group has engaged numerous caver colleagues in the research and survey of underground shelters throughout the projects conducted since 2014, some of which are featured in this book. The utmost effort in the execution of these initiatives and this book is attributable to them, and we express our deepest thanks to all our caver friends.

Contents

1 Introduction ... 1
1.1 Introduction ... 1
References ... 2

2 Cappadocia ... 3
2.1 Introduction ... 3
2.2 Geology ... 5
2.3 Historical Background ... 10
2.3.1 Historical Timeline for Cappadocia ... 14
2.4 Rock-Cut Architecture ... 15
2.5 Previous Researches in Cappadocia ... 18
References ... 22

3 Underground Shelters of Cappadocia ... 25
3.1 Introduction ... 25
3.2 Underground Shelters ... 27
3.2.1 Underground City vs Underground Shelter ... 28
3.2.2 Typology of Underground Shelters ... 29
3.3 Dating the Underground Shelters ... 31
3.4 Mechanisms and Architectural Features of Underground Shelters ... 35
3.4.1 Millstone Doors ... 36
3.4.2 Operation Rooms ... 38
3.4.3 Barriers and Shield Doors ... 39
3.4.4 Tunnels ... 39
3.4.5 Traps ... 40
3.4.6 Modules and Systems ... 40
3.4.7 Isolating Rooms and Isolated Areas ... 43
3.5 Excavation Techniques of Underground Shelters ... 43
3.5.1 Excavation on Opposing Fronts ... 43
3.5.2 Excavation Phases ... 45
3.5.3 Niches for Lamps ... 46
3.5.4 Pilot Holes ... 46
3.6 List of Underground Shelters ... 47
References ... 51

4 50 Significant Underground Shelters ... 55
4.1 Acıgöl Underground Shelter No 1 ... 55
4.2 Acıgöl Underground Shelter No 3 (Topada) ... 56
4.3 Ağırnas Underground Shelter ... 57
4.4 Ali Dağı Underground Shelter ... 59
4.5 Aşağı Hasinli Underground Shelter ... 60
4.6 Belağası Underground Shelter No 1 ... 61
4.7 Belağası Underground Shelter No 2 ... 62

4.8 Belha (Yallı Damı) Underground Shelter 64
4.9 Büyük Bürüngüz Underground Shelter 65
4.10 Çıralı Sinkhole Underground Shelter 66
4.11 Çomaklı Underground Shelter 67
4.12 Danalık Underground Shelter 69
4.13 Derinkuyu Underground Shelter 70
4.14 Doğala Underground Shelter 73
4.15 Doğanlı Underground Shelter 74
4.16 Durmuş Evi (Derinkuyu 4) Underground Shelter 75
4.17 Emiruşağı Underground Shelter 76
4.18 Eski Pusatlı Underground Shelter 77
4.19 Fatinler Underground Shelter 79
4.20 Filiktepe Underground Shelter 80
4.21 Golgoli Underground Shelter No 1 83
4.22 Golgoli Underground Shelter No 2 84
4.23 Gomeda Underground Shelter 85
4.24 Göstesin (Ovaören) Underground Shelter 87
4.25 Güzelöz Underground Shelter 88
4.26 Güzelyurt (Gelveri) Underground Shelter 89
4.27 Karşı Kilise Underground Shelter No 1 91
4.28 Karşı Kilise Underground Shelter No 2 93
4.29 Kaymaklı Underground Shelter 94
4.30 Kemer Ören Underground Shelter 95
4.31 Kırlangıç Valley Underground Shelter 96
4.32 Kirkor Underground Shelter 98
4.33 Kömür Underground Shelter 99
4.34 Kuşcağız Underground Shelter 101
4.35 Mazı Underground Shelter 103
4.36 Meskendir Underground Shelter 103
4.37 Mimar Sinan Underground Shelter 106
4.38 Mucur Underground Shelter 108
4.39 Örentepe Underground Shelter 109
4.40 Ötedere Valley Underground Shelter No 1 112
4.41 Ötedere Valley Underground Shelter No 2 113
4.42 Özkonak Underground Shelter 114
4.43 Özlüce (Zile) Underground Shelter 115
4.44 Pazarören Underground Shelter 116
4.45 Sivasa Underground Shelter 118
4.46 St. Mercurius Underground Shelter 120
4.47 Şahmelik Valley Underground Shelter No 1 124
4.48 Talasıra Underground Shelter 125
4.49 Tatlarin Underground Shelter 127
4.50 Zemi Underground Shelter 128
References 130

5 Conclusion 133
References 135

Glossary 137

Index 141

About the Authors

Ali Yamaç is a Turkish speleologist. He was born in 1957 and has been caving since 1976. During these years he has surveyed and explored hundreds of caves, some of which are the longest and deepest caves in Turkey. He was the president of the Speleological Federation of Turkey in 2007–2009 and the president during the foundation of three different speleological organizations. He was the leader of the team that prepared the cave inventory of Turkey. In addition to his natural cave explorations, he has explored and surveyed numerous rock dwellings and underground cities in different regions of Turkey as part of his artificial cave projects. Beginning with surveys of the underground structures of Hagia Sophia and Topkapi Palace in Istanbul, his artificial cave projects continue with underground structure inventory projects in the provinces of Kayseri and Gaziantep. Both projects have been underway for more than ten years and both have been accepted to the UNESCO World Heritage Tentative List. At the request of the directors of the Nevşehir and Aksaray Archaeological Museums, he has surveyed the underground shelters of Derinkuyu, Mazıkoy, Mazıkoy 2, Tatlarin, Avanos, Golgoli, and St. Mercurius; he has also investigated rock-cut settlements and churches on the Euphrates River cliffs as part of the Euphrates River Cliff Settlements Project; and he has surveyed several settlements as part of the Kura River—Ardahan Region Survey.

Ezgi Tok is a Turkish speleologist. Her research focuses on microbial life and related components of cave environments. Since 2016, she has been exploring several karst caves in Turkey to determine microbial diversity, ecosystem functioning of microbial populations, and their relationship with psychochemical conditions in the cave ecosystem. Her research also investigates the influence of surface atmospheric conditions on the subsurface microclimate. Her current project is the first cave microbiology study in Turkey comparing caves from different climatic regions with different environmental conditions. She has been caving since 2009 and has participated in several national and international speleological expeditions in Turkey and abroad, such as the Lion of Crete expedition, the 2nd deepest cave in Greece, and the Keş and Çem cave expeditions, the 6th and 10th deepest caves in Turkey. She had surveyed and mapped several caves of Turkey, among which İnsuyu Cave, which is the longest and Kocain Cave, which is the largest chamber of Turkey are particularly important. In addition to these works, she has also participated in artificial cave projects. She has surveyed and mapped several underground shelters in Cappadocia and dozens of rock-cut structures as part of underground structure inventory projects in the provinces of Kayseri and Gaziantep.

List of Figures

Fig. 2.1 Map of Roman Empire provinces of Asia Minor and Syria Superior during Strabo (first century AD) (Jones, 1929) 4
Fig. 2.2 Map of Central Anatolian Volcanic Province (CAVP), adapted from Toprak and Göncüoğlu (1993) 6
Fig. 2.3 Hasandağ (photo E. Soner) 7
Fig. 2.4 Wall painting from Çatalhöyük site, dated ca. 6600 BC. It has been interpreted as one of the oldest known maps showing the village and eruption of the Hasandağ volcano (Mellaart, 1967) 8
Fig. 2.5 Erciyes Mountain (photo A.E. Keskin) 8
Fig. 2.6 Simplified active tectonic map of Central Anatolia. Red triangles indicate volcanoes, adapted from (Yıldırım et al., 2016) AT: Aksu Thrust, CAFZ: Central Anatolia Fault Zone, EAFZ: East Anatolia Fault Zone, IEFZ: İnönü-Eskişehir Fault Zone, MFZ: Malatya Fault Zone, NAFZ: North Anatolian Fault Zone, OFZ: Osmaniye Fault Zone, TGFZ: Tuz Gölü Fault Zone 9
Fig. 2.7 Aerial photo of Aşıklı Höyük (Google Earth Pro) 10
Fig. 2.8 Kültepe (Kanis-Karum) (photo A.E. Keskin) 11
Fig. 2.9 Coin of Tiberius minted in Caesarea—Cappadocia AD 14–37. Silver drachm with Mount Argaeus on the back side (www.cngcoins.com) 12
Fig. 2.10 Coin of Hadrian minted in Caesarea—Cappadocia AD 117–138. Silver didrachm with Mount Argaeus on the back side (www.cngcoins.com) 12
Fig. 2.11 The province of antique Cappadocia in Kiepert's map (Kiepert, 1903) 13
Fig. 2.12 Locations of some dense cave settlement areas mentioned in this chapter. Signs do not indicate a specific point, but a larger area in that region: 1 Cappadocia, 2 Koramaz Valley, 3 Phrygian Highlands, 4 Gödet Village, 5 Gaziantep, 6 Halfeti, 7 Hasankeyf, 8 Ahlat, 9 Ani, 10 Kura River (Compiled by A. Yamaç, Google maps) 17
Fig. 2.13 Ayşepınar—Kayseri, Late Hellenistic rock-cut tombs (photo A. Yamaç) 18
Fig. 2.14 This engraving by Lucas is the first visual depiction of Cappadocia (Lucas, 1712) 19
Fig. 2.15 Facade of a rock-cut structure in an engraving by Texier. This structure, which does not exist today, is thought to be the courtyard façade of the Bezirhane in Avcılar, near Ürgüp (Kalas, 2004; Texier, 1839) 20
Fig. 2.16 An engraving of Uçhisar—Cappadocia (Hamilton, 1842) 20
Fig. 2.17 One of the earliest photographs of Uçhisar—Cappadocia by J.H. Haynes dated 1884 (Ousterhout, 2011) 21

Fig. 2.18 A fresco from the Kılıçlar Church, Göreme (Jerphanion, 1925–1942) (drawing Ernest Mamboury 1912). 21
Fig. 3.1 3D sketch of Kaymaklı underground shelter by Mahmut Akok (Akok & Gürçay, 1965) 26
Fig. 3.2 Örentepe shelter plan (Urban, 1969) 26
Fig. 3.3 The formation stages of an underground city (drawn A. Yamaç) 28
Fig. 3.4 Arab dominance in the region during tenth century AD (base map HGM—elaboration A. Yamaç) 32
Fig. 3.5 Topada monument (photo E. Soner) 33
Fig. 3.6 Karaburna monument (photo E. Soner). 33
Fig. 3.7 Map of Phrygian and Assyrian kingdoms around 670 BC (Wikimedia—elaboration A. Yamaç). 34
Fig. 3.8 Cross engraving on the wall of Golgoli underground shelter no 1 (photo A. Yamaç) 35
Fig. 3.9 Church inside the Ağırnas underground shelter (photo R. Straub) 35
Fig. 3.10 One of the earliest photos of a millstone door in an unknown shelter, probably Kaymaklı, 1935 (photo J. D. Whiting). 37
Fig. 3.11 A millstone door abandoned during the transportation in Ötedere shelter no 2 (photo A.E. Keskin) 37
Fig. 3.12 An unfinished millstone door extracted from the main rock in Kayakapı Orta Church / Ürgüp (photo B. Yazlık) 37
Fig. 3.13 A colossal millstone door in the Dereler cliff settlement—Ürgüp (photo A. Yamaç) 38
Fig. 3.14 Basic plan of an operation room (Bixio, 2012) 38
Fig. 3.15 Millstone door at the end of a tight tunnel at Büyük Bürüngüz underground shelter–Kayseri (photo A.E. Keskin) 39
Fig. 3.16 An ascending trap (Bixio, 2012) 40
Fig. 3.17 A shaft/trap with a depth of 5 m in an underground shelter in Ürgüp (photo B. Yazlık). 41
Fig. 3.18 A simplified sketch showing a vertical shelter and its shaft/trap (Bixio, 2012) 41
Fig. 3.19 Sketch of a module (Bixio, 2012) 42
Fig. 3.20 Sketch of a system formed by the combination of three different modules (Bixio, 2012). 42
Fig. 3.21 Typical opposing fronts excavation and errors in Güvercinlik water tunnel (drawn E. Gilli) (Gilli & Yamaç, 2015) 44
Fig. 3.22 Semicircular axe marks on the wall of a shelter in Ürgüp (photo A. Yamaç) 45
Fig. 3.23 Tunnel excavation on opposing fronts with shafts (Bixio, 2012) 45
Fig. 3.24 An antique oil lamp in a rock-cut niche (photo A. Yamaç) 46
Fig. 3.25 Underground shelters at the farthest points of Cappadocia; 1. Çıralı 2. Pazarören 3. Çeşka 4. Fertek (after Google Maps—elaboration A. Yamaç). 48
Fig. 3.26 Underground shelter density map (drawn A. Yamaç) 48
Fig. 4.1 Acıgöl Underground Shelter No 1 plan (Bixio, 2012). 56
Fig. 4.2 Acıgöl Underground Shelter No 3 (Topada) plan (Bixio, 2012). 56
Fig. 4.3 Ağırnas Underground Shelter plan (drawn E. Tok, Ç. Çankırılı) 58
Fig. 4.4 A millstone door and operation room in Ağırnas Underground Shelter (photo R. Straub). 59
Fig. 4.5 Ali Dağı Underground Shelter plan (drawn A. Yamaç). 60
Fig. 4.6 Cistern inside the Ali Dağı Underground Shelter (photo A. Yamaç). 60
Fig. 4.7 Plans of sectors 1 and 2 of Aşağı Hasinli Underground Shelter (Bixio, 2012) 61
Fig. 4.8 Belağası Underground Shelter No 1 plan (drawn E. Tok). 62
Fig. 4.9 Belağası Underground Shelter No 2 plan (drawn A.E. Keskin, E. Tok) 63

Fig. 4.10 Belağası Underground Shelter No 2 (photo A.E. Keskin) 63
Fig. 4.11 Belha (Yallı Damı) Underground Shelter (Bixio, 2012) 64
Fig. 4.12 Separate tunnels in Büyük Bürüngüz Underground Shelter just after the first entrance (photo A.E. Keskin) 65
Fig. 4.13 Büyük Bürüngüz Underground Shelter plan (drawn A. Yamaç, B. Yazlık) 67
Fig. 4.14 Cliff settlements in the Çıralı Sinkhole, Konya Province (photo A.E. Keskin) 68
Fig. 4.15 Çıralı Underground Shelter (drawn M. Leontiev) 68
Fig. 4.16 Çomaklı Underground Shelter plan (drawn K. Kendirciler, T. Değirmenci) 69
Fig. 4.17 Danalık Underground Shelter plan (drawn A. Yamaç) 70
Fig. 4.18 Derinkuyu Underground Shelter (photo M. Gülyaz). 72
Fig. 4.19 Derinkuyu Underground Shelter sketch plan (Bixio, 2012) 72
Fig. 4.20 Doğala Underground Shelter System DO1 plan (Bixio, 2012) 73
Fig. 4.21 Doğanlı Underground Shelter main chamber (photo A. E. Keskin) 74
Fig. 4.22 Doğanlı Underground Shelter plan (drawn E. Tok, A. Yamaç) 75
Fig. 4.23 Durmuş Evi Underground Shelter plan (Bixio, 2012). 77
Fig. 4.24 Emiruşağı Underground Shelter (photo A.E. Keskin) 78
Fig. 4.25 Emiruşağı Underground Shelter plan (drawn A. Yamaç, E. Gilli). 78
Fig. 4.26 Entrance of Eski Pusatlı Underground Shelter (photo A.E. Keskin). 79
Fig. 4.27 Eski Pusatlı Underground Shelter (drawn A. Yamaç) 80
Fig. 4.28 Fatinler Underground Shelter (drawn A. Yamaç). 81
Fig. 4.29 Fatinler Underground Shelter (photo A.E. Keskin) 81
Fig. 4.30 Filiktepe Underground Shelter, plan of sectors F1, F2 and F3 (Bixio, 2012) 82
Fig. 4.31 Filiktepe Underground Shelter, plan of sectors F4–F9 (Bixio, 2012) 83
Fig. 4.32 Golgoli Underground Shelter No 1 (drawn E. Gilli, A. Yamaç) 84
Fig. 4.33 Millstone door in the upper room of Golgoli Underground Shelter No 1 (photo E. Gilli) 85
Fig. 4.34 Golgoli Underground Shelter No 2 plan (drawn E. Gilli, A. Yamaç) 86
Fig. 4.35 Gomeda Underground Shelter plan (drawn N. Moldavskaya) 87
Fig. 4.36 Göstesin (Ovaören) Underground Shelter (Bixio, 2012). 89
Fig. 4.37 Main passage of Güzelöz Underground Shelter (photo A. E. Keskin) 90
Fig. 4.38 Güzelöz Underground Shelter plan (drawn E. Tok) 91
Fig. 4.39 Güzelyurt Underground Shelter plan (Bixio, 2012) 92
Fig. 4.40 Karşı Kilise Underground Shelter No 1 plan (Bixio, 2012) 93
Fig. 4.41 Karşı Kilise Underground Shelter No 2 plan (Bixio, 2012) 95
Fig. 4.42 Kaymaklı Underground Shelter (photo A. E. Keskin). 96
Fig. 4.43 Kaymaklı Underground Shelter plan (Bixio, 2012). 97
Fig. 4.44 Kemer Ören Underground Shelter plan (Bixio, 2012). 98
Fig. 4.45 Entrance of Kırlangıç Valley Underground Shelter (photo A. E. Keskin) 99
Fig. 4.46 Kırlangıç Valley Underground Shelter plan (drawn E. Tok) 100
Fig. 4.47 Kirkor Underground Shelter (photo A. E. Keskin) 101
Fig. 4.48 Kirkor Underground Shelter plan (drawn A. Yamaç) 102
Fig. 4.49 Kömür Underground Shelter (photo A. E. Keskin) 103
Fig. 4.50 Kömür Underground Shelter plan (drawn E. Tok, E. Gilli). 104
Fig. 4.51 Kuşcağız Underground Shelter plan (drawn A. Yamaç, E. Tok, E. Gilli) 105
Fig. 4.52 Kuşcağız Underground Shelter (photo A. E. Keskin) 106
Fig. 4.53 Mazı Underground Shelter (drawn E. Tok) 107
Fig. 4.54 Meskendir Underground Shelter plan (Bobrovskyy & Grek, 2015) 108
Fig. 4.55 Mimar Sinan Underground Shelter (photo Ç. Çankırılı). 109

Fig. 4.56 Mimar Sinan Underground Shelter (drawn E. Tok, E. Gilli, Ç. Çankırılı, A. Yamaç) 110
Fig. 4.57 Mucur Underground Shelter (Bixio, 2012) 111
Fig. 4.58 Örentepe Underground Shelter plan (Bixio, 2012) 112
Fig. 4.59 Ötedere Valley Underground Shelter No 1 (photo A.E. Keskin) 112
Fig. 4.60 Ötedere Valley Underground Shelter No 1 plan (drawn A. Yamaç) 113
Fig. 4.61 Ötedere Valley Underground Shelter No 2 plan (drawn A. Yamaç) 114
Fig. 4.62 ÖzkonakUnderground Shelter (photo M. Gülyaz) 115
Fig. 4.63 ÖzkonakUnderground Shelter plan (Bixio, 2012) 116
Fig. 4.64 Özlüce (Zile) Underground Shelter plan (Bixio, 2012) 117
Fig. 4.65 Pazarören Underground Shelter (drawn H.İ. Kala, A. Yamaç) 118
Fig. 4.66 Sivasa Underground Shelter System S1 plan (Bixio, 2012) 119
Fig. 4.67 Sivasa Underground Shelter plans of systems S1 to S11 (Bixio, 2012) 121
Fig. 4.68 St. Mercurius Underground Shelter 1. and 2. floors (drawn E. Tok) 122
Fig. 4.69 St. Mercurius Underground Shelter 3. floor (drawn E. Tok) 123
Fig. 4.70 Church and the graves in St. Mercurius Underground Shelter (photo E. Soner) 123
Fig. 4.71 Şahmelik Valley East Wall (photo A. Yamaç) 124
Fig. 4.72 Şahmelik Valley Underground Shelter No 1 plan (drawn A. Yamaç) 125
Fig. 4.73 Talasıra underground shelter plan (drawn A. Yamaç) 126
Fig. 4.74 Entrance of Talasıra Underground Shelter with a millstone door (photo A.E. Keskin) 127
Fig. 4.75 Tatlarin Underground Shelter (photo M. Gülyaz) 128
Fig. 4.76 Tatlarin Underground Shelter plan (drawn E. Tok) 129
Fig. 4.77 Zemi Underground Shelter plan and profile (drawn O. Klimishina, N. Moldavska) 130

Introduction

1

Abstract

Although subterranean defense shelters may be found in many different regions of the world, the province of Cappadocia, which is located in Central Anatolia, is particularly renowned for these structures. The underground shelters of Cappadocia, which was an important region for both Rome and Byzantium, have been the subject of a great number of publications. However, most of these publications provide basic information and are insufficient to address the history, types, or construction techniques of these structures. Additionally, there is a lack of a comprehensive database documenting these shelters of Cappadocia. This work attempts to fill these deficiencies.

Keywords

Underground shelter · Cappadocia

1.1 Introduction

Somewhere in the vast steppe of Central Anatolia, you come across a hole dug in the ground. You bend down and look. In the depths of the pit is a large, round stone, which has been partially buried over the past centuries. This is a millstone door, the door of a shelter that was dug hundreds of years ago.

The same phenomenon is observed not only in the pits in the steppes, but also in the villages of Central Anatolia. There is a tunnel in a room of an old house. You progress through this tunnel, sometimes walking and sometimes crawling, and you will eventually encounter the aforementioned millstone door once more. The tunnel extends beneath the neighbor's residence, subsequently traversing another dwelling. If the tunnels have not collapsed or become clogged over the past centuries, they would extend throughout an entire village.

The date and the identity of the builders of these structures and the doors that protect them are unknown. Over the centuries, these enigmatic constructions have been the subject of numerous legends, science fiction stories, and conspiracy theories. Local residents, even treasure hunters, are reluctant to go through these tunnels. In addition to the irrational reasons for these fears, which are based on superstition or demons, there is also another very logical reason, such as the possibility of becoming lost in these labyrinthine tunnels and being unable to find the exit again.

These are the underground shelters of Cappadocia, and it is understandable that treasure hunters are apprehensive. Even the shelters that have been cleaned, illuminated, and made accessible to tourists are like labyrinths. It is difficult to imagine attempting to navigate them by crawling through a semi-dark tunnel and trying to find your way. Moreover, the distances involved are considerable, extending to hundreds of meters. As you will see in later chapters, these distances reach 2000 m in extreme underground shelters such as Derinkuyu or Büyük Bürüngüz.

People have been using the underground for thousands of years to hide and protect themselves. In addition to the difficulty of detecting the entrances of these shelters, numerous rooms connected by intricate tunnels filled with traps and doors that are almost impossible to open from the outside have turned these structures into defense centers indispensable against all types of attacks throughout history.

The rich tapestry of history unfolds beneath the surface, where humans, for millennia, have sought refuge and defense in the clandestine world beneath our feet. Subterranean shelters, shrouded in mystery and strategic brilliance, have stood as silent guardians against countless attacks throughout the ages. These architectural marvels, interconnected by labyrinthine tunnels laden with traps and impregnable doors, weave a narrative of resilience spanning continents.

Therefore, in numerous different geographies from northern France to Vietnam or Afghanistan, there are countless underground defense structures today and these

A. Yamaç and E. Tok, *Underground Shelters of Cappadocia, Turkey*, Advances in Karst Science,
https://doi.org/10.1007/978-3-032-16095-9_1

relevant structures have been used from thousands of years ago till recent history (Triolet & Triolet, 1996, 2013).

Though they have been used for centuries in numerous different countries, the first place that comes to mind when it comes to "underground defense structures" is always Cappadocia. The two main reasons are the excessive number of underground shelters in this region and the enormous size of some of these structures. Large or small, virtually every Cappadocian village has at least one, if not more, subterranean shelter. Shelters are widely distributed across Kayseri and Nevşehir provinces, as well as extending to Yozgat, Niğde, Aksaray, and Kırşehir provinces (Bixio, 2012). The excavation of these structures is frequently attributed to Christians during the turbulent seventh to eleventh centuries, who constructed these subterranean refuges as a defense against Arab invasions.

However, the lack of scientific research on Cappadocian underground shelters makes it necessary to conduct additional research in order to uncover the delicate features of these ancient defensive systems.

Cappadocia, in central Anatolia, preserves the tradition of cave dwelling to this day. It's a region famous for its enchanting fairy chimneys, rock-cut churches, picturesque valleys, and breathtaking natural scenery. Attracting thousands of tourists each year, it is one of Turkey's most distinctive destinations.

At the heart of Cappadocia is Göreme National Park, home to a collection of eighth- to eleventh-century rock-cut churches, which was accepted as a UNESCO World Heritage Site in 1985. Within this compact area, a wealth of intricately decorated rock-cut churches, dwellings, and underground shelters tell the story of a troglodyte civilization that flourished here for centuries, if not millennia.

Although this historical past of the region has been known for over a century, the primary emphasis of research and study in Cappadocia has been on frescoed rock-cut churches. Apart from a limited number of articles on rock-cut structures other than churches, there is a lack of significant scientific research.

For example, most of the literature on cliff settlements in valleys is written at a superficial level, resembling tourist guides. However, these rock-cut villages, utilized for centuries, are the most significant exemplars of the region's troglodyte civilization and way of life. Despite the historical importance, there has been no comprehensive study of the rock-cut settlements of Zelve, Çavuşin, and even the Peristrema Valley.

Likewise, although the pace of research has accelerated in recent years, additional scientific inquiry is necessary to achieve a more thorough comprehension of the dating, origins, defense, and excavation techniques utilized in subterranean shelters.

In conclusion, Cappadocia's underground shelters represent a fascinating intersection of history, architecture, and defense. As we strive to unravel the mysteries of these ancient marvels, we are confronted with a complex tapestry of human ingenuity and adaptation, spanning millennia of civilization.

This book, which is essentially an inventory of the underground shelters in Cappadocia, examines the hundreds of defense structures that have been found within the region so far from a variety of angles, including dating, architecture, excavation techniques, and typologies. In addition, ongoing academic debates on these issues will be analyzed in detail in different chapters of the book.

After this short introduction, Chap. 2 is structured with four distinct sections. After a section on the geological events that led to the formation of all these underground and rock-cut structures in Cappadocia and that extend from millions of years ago to the present day, another section summarizes the historical background of the region. The final two sections provide an overview of the architectural characteristics of all the rock-cut buildings in the area, including the shelters, as well as the earlier studies that were conducted there.

The core of this book is in Chap. 3. In this chapter, typology, distribution, the excavation techniques of underground shelters, the defensive devices used in these structures, and dating debates are discussed in detail. Then, in Sect. 3.6, a full list of underground shelters identified to date is presented. 50 important underground shelters, selected from this list, researched and surveyed, are discussed in detail in Chap. 4. All 50 shelters in this chapter have been scientifically researched, mapped, and documented by different teams.

Chapter 5, which examines the endeavors necessary to maintain the underground shelters, each exceeding a millennium in age, for future generations, serves as a finale while embodying the primary objective of this entire book.

These structures in Cappadocia, a region of profound cultural and historical significance in Anatolia, have been subject to scientific neglect for an extended period. Consequently, they are gradually disappearing due to a lack of interest and proper conservation efforts.

References

Bixio, R. (Ed.). (2012). *Cappadocia: Records of the underground sites*. Archaeopress.

Triolet, J., & Triolet, L. (1996). Deux Conceptions du Grand Souterrain-Refuge Villageois Hiermont (Picardie-France) et Sivasa (Cappadoce-Turquie). *Subterranean, 97* (Archeologie des Mondes Souterrains).

Triolet, J., & Triolet, L. (2013). Underground refuges and war tunnels (France, Cappadocia, Afghanistan, Vietnam and Lebanon). *Opera Ipogea, 1*.

Cappadocia

2

Abstract

Thousands of articles and hundreds of books have been written about the geology, history, and rock-cut architecture of Cappadocia. This chapter will therefore only provide a concise, broad summary of the aforementioned topics. For readers seeking comprehensive knowledge on these topics, the references serve as a guide. The initial section will present a concise review of the geological and geomorphological characteristics of Cappadocia, focusing on volcanic and tectonic processes. A brief review of the region's history will follow. The final section includes a description of the rock-cut structures located throughout Turkey, beyond the Cappadocia region, as well as analyses of those monuments. Following that, a comprehensive overview of prior and current research on rock-cut architecture in the Cappadocia region will be provided.

Keywords

Cappadocia geology · Rock-cut structures · Artificial cavity

2.1 Introduction

Cappadocia is one of the rare regions in the world whose name has remained nearly unchanged for approximately 2500 years. The earliest recorded mention of the name appears in the works of the ancient Greek historian Herodotus (c. 484–c. 425 BC), who referenced the area in his historical writings (Herodotus, 1920). The precise etymology of the name "Cappadocia" has been a subject of scholarly debate for many years. In its Greek form, Καππαδοκία (Kappadokía), the name is believed to have been derived from the Old Persian term "Katpatuka." Interestingly, at some point in history, this word was incorrectly translated as "The land of beautiful horses," a misinterpretation that has persisted in popular usage over the years. This inaccurate translation likely gained widespread acceptance due to its appealing and romanticized portrayal of the region's historical significance.

A more linguistically rigorous explanation has been proposed by the linguist Ilya Yakubovich (2014). According to his well-substantiated argument, the term "Katpatuka" is derived from the Hittite language, where "Katta-peda" was documented in ancient records.

In its Luvian form, "Katpadda" translates to "Low Country," a name that is geographically appropriate. This designation is thought to reference Hattusa, the capital of the Hittite Empire, which is located north of Cappadocia at a significantly higher elevation. Thus, the term "Low Country" would have served to distinguish Cappadocia in relation to the Hittite heartland.

In order to understand Cappadocia, one must first understand the geographical standing of the region as an area of osmosis between eastern and western Anatolia, which fostered settlement since ancient times. Set at the center of the Anatolian peninsula (from ancient Greek *anatole*/*anatellein,* i.e., "the land of the rising sun"), Cappadocia shows strong cultural contact with Middle Eastern regions. Although the term "Cappadocia" is now commonly associated with a relatively small region recognized for its tourism appeal, historically, it encompassed a significantly larger territory. During the Roman and Byzantine periods, Cappadocia was one of the most expansive provinces within both empires (Fig. 2.1). A more detailed discussion of this historical evolution will be provided in Sect. 2.3.

Cappadocia holds immense global significance, not only due to its distinctive relief, which harmoniously reflects the diverse features of geomorphology, but also because of its extensive cultural and archaeological heritage. This vast plateau, situated in the southeastern part of Central Anatolia, is characterized by a remarkable landscape

A. Yamaç and E. Tok, *Underground Shelters of Cappadocia, Turkey*, Advances in Karst Science,
https://doi.org/10.1007/978-3-032-16095-9_2

Fig. 2.1 Map of Roman Empire provinces of Asia Minor and Syria Superior during Strabo (first century AD) (Jones, 1929)

shaped by volcanic activity and subsequent erosion. In geological and geographical discourse, it is frequently referred to as the Central Anatolian Volcanic Province (CAVP) or the Cappadocian Volcanic Plateau. The region's natural topography is unparalleled, featuring an extraordinary array of landforms, including fairy chimneys, expansive plains, undulating hills, stunning valleys, and towering volcanic mountains. These formations create a striking interplay of colors and shapes, contributing to Cappadocia's unique visual identity.

Beyond its natural wonders, Cappadocia is also distinguished by its human-made features. The region, with an area of approximately 20,000 km^2, contains an extensive network of troglodyte (cave dwelling) settlements, intricately painted rock-cut churches, and vast underground shelters. These structures provide a unique testament to the ingenuity and resilience of the people who inhabited this area over the centuries.

A significant portion of the Cappadocian landscape is described as volcanic terrain, based on the rock unit within the geological formation. These geological formations, which include rock units primarily composed of tuff, can reach thicknesses of up to 300 m in certain areas. Due to their relatively soft consistency, these formations are highly susceptible to both natural erosion and human excavation. Over the centuries, the inhabitants of the region took advantage of this characteristic, creating an extensive subterranean infrastructure within the malleable volcanic material.

The Cappadocians, who skillfully excavated structures in these volcanic rocks, established a troglodyte way of life that persisted for hundreds of years. Within this landscape, they constructed not only residential dwellings but also functional spaces such as barns, storage facilities, water tunnels, cisterns, dovecotes, and apiaries. The ability to shape such a diverse range of structures directly from the rock highlights the resourcefulness of these communities. The ability to carve out such a wide variety of structures from the rock highlights the resourcefulness of the people who lived in Cappadocia throughout history.

Among the most significant and intriguing elements of this rock-hewn civilization are its underground shelters. These extensive subterranean networks, likely constructed between the seventh and eleventh centuries AD, served as defensive refuges during prolonged periods of Arab incursions. The persistent threat of invasions necessitated the development of increasingly sophisticated underground cities, equipped with millstone doors, ventilation systems, and concealed entrances to ensure the safety of their inhabitants.

While many scholars associate Cappadocia's troglodyte civilization primarily with the Byzantine period due to the prevalence of rock-hewn churches and defensive underground structures, some researchers, including Spiro Kostof, have challenged this assumption. Kostof and his colleagues argue that "there is no reason to disbelieve that the practice was more ancient than Christianity" (Kostof, 1972). This perspective is supported by the presence of numerous Roman rock-cut tombs in the region, which suggest that the tradition of carving structures into the rock predates the spread of Christianity in Cappadocia.

Over time, as social and political conditions evolved, the troglodyte civilization gradually declined. As new building techniques and construction materials became more widely available, people began to abandon their subterranean dwellings in favor of more conventional forms of architecture. Despite this shift, the underground shelters of Cappadocia remain as enduring symbols of the region's past. Today, these subterranean complexes continue to provide invaluable insights into the lives of their former inhabitants, attracting both scholars and tourists interested in their remarkable history.

Unfortunately, for more than a century, academic research on Cappadocia has largely focused on its frescoed rock-cut churches, often neglecting the broader scope of its troglodyte civilization. Only in recent years has serious scholarly attention expanded to encompass the study of underground settlements and other rock-hewn structures.

Cappadocia represents a unique and invaluable "laboratory" for examining the formation and development of troglodyte civilizations. The sheer volume and variety of rock-cut structures offer an exceptional opportunity to analyze the architectural adaptations and innovations of past societies. In the early 1990s, the National Commission for Artificial Cavities of the Italian Speleological Society initiated systematic research efforts in the region. These investigations, followed by the publication of numerous articles and books by the research team, have played a crucial role in uncovering previously unexplored aspects of Cappadocia's subterranean heritage.

Through ongoing archaeological research and conservation efforts, Cappadocia's hidden history continues to be revealed. The region remains an extraordinary testament to human adaptability, ingenuity, and resilience, solidifying its status as one of the most fascinating cultural and historical landscapes in the world.

2.2 Geology

The region, which is defined as the Central Anatolian Volcanic Province (CAVP) in the geological literature, is one of the most important volcanic areas of Turkey. The area extends approximately in the NE-SW direction, the long axis is 200 km and its width is around 100 km, covering around 20.000 km^2 (Le Pennec et al., 1994) (Fig. 2.2).

The formation of CAVP began with the converging of the Arabian and Eurasian plates in the Middle Miocene period, around 16–11.6 million years ago, and developed the post clash regimes after the Upper Miocene period. Because of this, many different types and origins of volcanic rocks were formed within the CAVP. The orogenic activity of the Taurides continued during the Miocene period causing deep fractures in the crystalline mountains in the north. These deeply located fractures caused a weakening of the crust followed by the generation of a chain of volcanic mountains at the heart of central Anatolia (Andolfato & Zucchi, 1971). This continental clash brought shortening and thickening together with the north–south compression regime that is still going on today in Eastern Anatolia. The thickness of the Eastern Anatolian Crust is about 45–50 km and at the end of this clash, the Anatolian block moved to the west through the two strike-slip faults. This movement that pulled apart the basin and reverses faults is the basic reason for the inner deformation within the Anatolian block (Alıcı Şen et al., 2004). Wide-scale volcanism was formed in Anatolia from the Neogene to Quaternary periods as a result of the convergence of plates and continental clashes. Around the Cappadocia region, notable volcanic peaks such as Erciyes, Develi, Hasan, Melendiz, Keçiboyunduran, and Göllü mountains bore witness to millennia of volcanic activity, shaping the landscape and leaving behind a legacy of majestic cones and rugged terrain (Aydar et al., 2012; Innocenti et al., 1975; Le Pennec et al., 2005).

It is known that the widespread volcanism in the area between Kayseri and Nevşehir generally started in the Late Miocene period, approximately between 11 and 7 million years.

Volcanic activity in the vicinity of Mount Erciyes commenced with andesitic lavas, gradually evolving to include tuffs and ignimbrites. These pyroclastic rocks, forming expansive plateaus, became prominent features of the landscape. The oldest volcanic rocks exposed and dated in ancient Cappadocia, which cover a large part of this region, are 10.1 million years old andesitic lavas, which are observed in the west of Erciyes Mountain and southeast of Ürgüp. The "Akköy ignimbrites" of the following

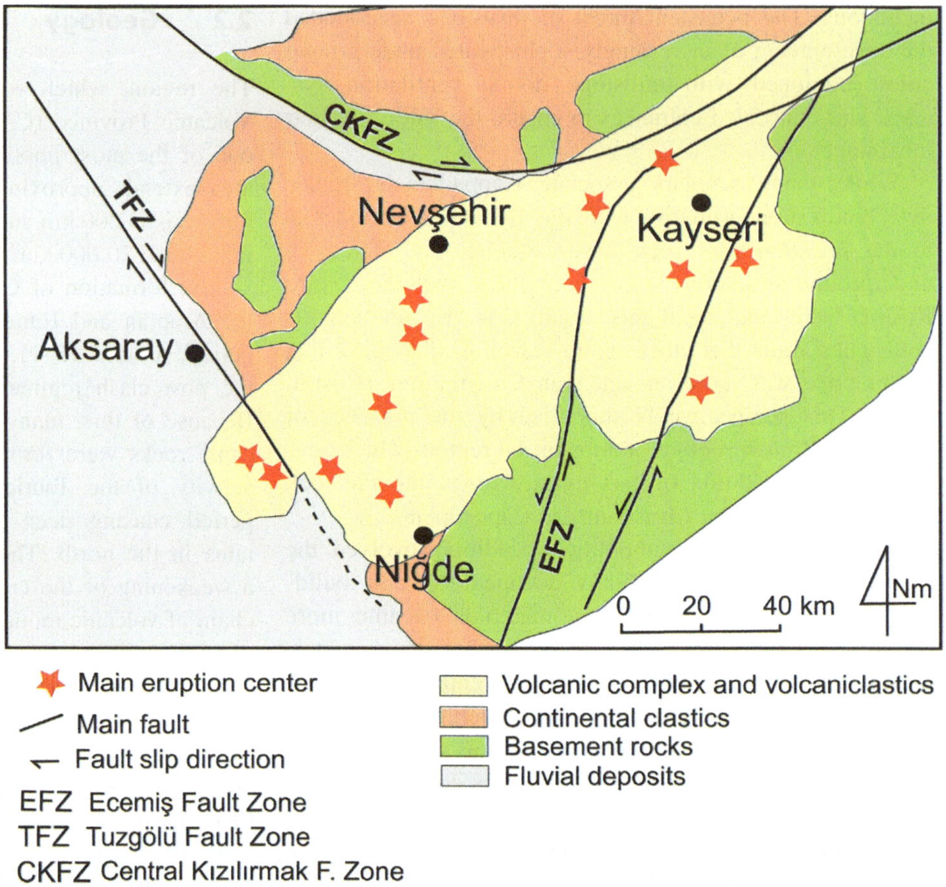

Fig. 2.2 Map of Central Anatolian Volcanic Province (CAVP), adapted from Toprak and Göncüoğlu (1993)

pyroclastic activity are dated to 8.5 million years, and the overlying "Velibaba Tepe ignimbrites," which are spread over an area of approximately 2.000 km^2, are dated to 2.7–3 million years (Şen et al., 2003).

In addition, the following different radiometric ages were also determined in the vicinity of Mount Erciyes (Ketin, 1995):

- Andesitic lava on the western slope of Erciyes Mountain: 300,000 years
- Dacitic lava on the southern slope of Erciyes Mountain: 900,000 years
- Ignimbrite near Erkilet: 2.8 million years
- Dacitic ignimbrite near İncesu: 3.0 million years

As a result of many different age determinations, it is understood that the activity of the main caldera of Erciyes Mountain continued until 150,000–200,000 years ago and after this period, the caldera was blocked (Alıcı Şen et al., 2004). But, cosmogenic datings show that several smaller lava flows occurred from monogenetic vents during the Early Holocene (ca. 10,000–8000 years ago).

The geomorphology of the CAVP is dominated by plateaus, cut in places by valleys, where fairy chimneys and man-made caves coexist. The Kizilirmak River to the north flows mostly within these volcanic and fluvio-lacustrine products. Channel deposits of this river are also often found unconformably overlying these ignimbrites, creating peculiar morphological features.

The strongly welded Kizilkaya ignimbrite is the most widespread unit in the CAVP. It covers most of the underlying deposits and forms a high plateau with flat topography (mesa) from the Soğanlı valley in the east to the Ihlara valley in the west. The plateau is cut by palaeo and more recent fluvial systems, forming valleys and deep gorges, some of which are several hundred meters deep.

The ignimbrites and lava flows in the CAVP were first described by Pasquarè, and the stratigraphy was further refined by numerical ages obtained using different dating techniques. Recently, Aydar et al. (2012) published Ar/Ar plagioclase eruption and Pb/U zircon crystallization ages, in which they refined the stratigraphy by defining a total of ten ignimbrite sequences following the terminology described by Le Pennec et al. (1994).

The stratigraphy of the Cappadocian ignimbrites, meticulously studied and refined over the years, revealed the complex history of volcanic eruptions in the region. The ignimbrite sequences, ranging from 10 million years to the Quaternary period, bear witness to a succession of volcanic events, each leaving its mark on the geological record.

The Cappadocian ignimbrites were mistakenly thought to be the products of dominant volcanoes, and the rhyolitic massive of the Göllüdağ volcanoes. In fact, the Late Miocene-Pliocene ignimbrites have their own eruption centers, leading to caldera collapse. Although the source areas are still debated, the Kavak and Sanmadentepe ignimbrites are thought to be related to the Çardak caldera, and the Zelve and Kumtepe ignimbrites to the Erdaş-Acıgöl area. Tahar ignimbrite probably originated east of Derinkuyu. The source area of Sofular ignimbrite was proposed by Le Pennec et al. (1994) to be below the Topuzdağ area. The source of the Kızılkaya and Cemilkoy ignimbrites is still a mystery, proposed to be in the Derinkuyu basin.

There are two Quaternary stratovolcanoes in Cappadocia: Erciyes and Hasandağ. Both volcanoes have their own evolutionary history.

Hasandağ is a double-peaked composite stratovolcano, culminating on a plateau one km above sea level (Fig. 2.3). Multiple evolutionary stages have been identified as Paleo-, Meso- and Neo-Hasandag through emplacement of extrusive domes and intermittent collapse events associated with ignimbrites. Limited geochronological data indicate emplacement of the oldest lavas at 7.21 Ma and ignimbrite emplacement during an early caldera collapse at 6.31 Ma, coincident with widespread Neogene ignimbrite volcanism in Cappadocia. Dome extrusion with associated block and ash- deposits and adventive monogenetic vent eruptions at the base are collectively attributed to the Neo-Hasandağ stage—the actual form of the volcano. Neo-Hasandağ, which consists of two summits, has numerous collapsed andesitic to rhyodacitic lava domes on its flanks, resulting in widespread pyroclastic deposits. The resulting soft block and ash-flow deposits are deeply eroded and intensively carved. Debris avalanche deposits outcrop to the north of the volcano, forming a typical hilly surface. Hasandağ is considered an active-subactive volcano. K/Ar ages show that volcanic activity occurred during the Holocene with an andesitic lava dome extrusion on the northern flank.

The British archaeologist James Mellaart excavated the Neolithic settlement of Çatalhöyük in the Konya Plain in the 1960s. One of the most striking discoveries during the excavation was a large number of wall paintings, which were photographed and sketched on site. One of these depicted an erupting volcano behind a town (Mellaart, 1967) (Fig. 2.4). If this interpretation is correct, the painting is the oldest depiction of a volcanic eruption and also the first graphic representation of an event or even a landscape in the world. Recent U-Th/He eruption ages show that the Hasandağ volcano erupted explosively and was eyewitnessed 9000 years ago (Schmitt et al., 2014).

The other Quaternary stratovolcano, Erciyes, is a huge, voluminous volcano with several monogenetic vents on its flanks (Fig. 2.5). Erciyes was well known in ancient times and its name probably derives from Mont Argaeos (Greek) or Argaeus (Latin), meaning "bright" or "white." Mount Argaeos must have made a deep impression on the people of Caeseria (Kayseri) as they usually used a sketch of Argaeos on the tails of Roman provincial coins, while the heads depicted the gods, emperors, or kings.

After extensive research in the Kayseri Region, in his article dated 1995, Prof. İhsan Ketin summarizes the geomorphological formation in the region as follows:

> There are 68 volcanic cones, large and small, with or without a crater, with diameters of 600 – 3,000 m around the 3,917 m high central cone of Mount Erciyes, which is the most prominent with its size and height among the extinct volcanoes of

Fig. 2.3 Hasandağ (photo E. Soner)

Fig. 2.4 Wall painting from Çatalhöyük site, dated ca. 6600 BC. It has been interpreted as one of the oldest known maps showing the village and eruption of the Hasandağ volcano (Mellaart, 1967)

Fig. 2.5 Erciyes Mountain (photo A.E. Keskin)

Central Anatolia. Thus, Mount Erciyes is not a single volcano, but a group of volcanoes. Located between Kayseri-Develi-İncesu, the Erciyes Mountain volcanic group covers an area of approximately 1,500 km^2 together with the Erkilet region. (Ketin, 1995)

A range of Ar40/Ar39 and Potassium-Argon dates have been obtained for the volcanic activities of Erciyes Mountain. The oldest volcanic activity at Erciyes is known as Koç Dağ, which forms the eastern slope of Erciyes. This complex erupted pyroxene andesite, 15 km of lava flows from the Topakkaya Tepe cone and 0.2 km^3 of dense rock equivalent of fall deposits and scoria from Kızıl Tepe. An age of 4.39 ± 0.28 million years was obtained at Koç Dağ. Overall, Koç Dağ was active between 4.4 and 2.9 million years ago.

Caldera-forming activity occurred in several eruptive phases, accompanied by pumice flows and ash fall. A first phase of activity produced plinian deposits that reach thicknesses of 22 m up to 21 km from the volcano and cover at least 3000 km of the surface. At least fifteen separate layers have been found. A second phase of activity produced pumice flows east-northeast of Koç Dağ, covering an area of 2100 km^2 and up to 8 m thick.

Eastern Cappadocia is home to the famous Cappadocian ignimbrites; one of these, the Velibaba Tepe ignimbrite, was associated with Erciyes volcano and is the last Cappadocian ignimbrite. This eruption 2.8 million years ago had a total volume of 52 km^3 and was preceded by a smaller Plinian eruption that covered an area of 1500 km^2 with pumice falls. The Velibaba Tepe ignimbrite extends to the east of Erciyes volcano, where it originated and filled in the previous topography. Its total volume is estimated at 146 km^3.

The Erciyes volcano was formed in two phases, starting with a basaltic andesitic lava flow on the southern slope 1.7 million years ago. This was followed by the andesitic lava flows on the western flank and then by many of the dacitic lava domes. The next phase of activity was explosive, with eruptions at the summit of Erciyes producing block and ash flows, pumice flows, and lava domes forming blocks 1.5–2 m in diameter. The deposits of this activity are found north and south of the summit of Erciyes and reach thicknesses of 18 m. Activity in the summit region probably ended before the Holocene.

On the other hand, a paragraph about Mt. Erciyes in Strabo's "Geographica," written in the first century AD, has been interpreted by some scholars as meaning that the mountain was also active at that time. In book 12, chapter 2, he wrote:

> At a little distance there are burning plains and pits full of fire to the extent of many stadia, so that the necessities of life are brought from afar.... In some places the ground is swampy, and at night flames burst from the earth. Those who are familiar with the country gather wood with care; but there is danger to others, and especially to cattle, who fall into these hidden fire pits. (Jones, 1929)

But no Late Holocene volcanic eruption of Mt. Erciyes has ever been proven. Therefore, Strabo must have seen methane vents rising to the surface and catching fire in the Sultan Sazlığı marshes, which lie in front of Mt. Erciyes (Aydar et al., 2019).

Dikkartin Dağ, Karagüllü Dağ, and Perikartın Dağ belong to the latest stages of volcanic activity at Erciyes and were formed on the rim of the former caldera. The three eruptions appear to have occurred within a short time of each other. Yılanlı Dağ, a lava dome in Kayseri, is also of Holocene age, having erupted 8900 ± 400 BP.

A 1 mm thick layer of tephra found in a drill core off the coast of Israel was dated to 8365 ± 65 BP in uncalibrated radiocarbon years. This tephra layer, based on its composition, appears to be associated with one of the three eruptions, most likely the Dikkartin eruption. Another layer of tephra found in Lebanon and dated to 8600 ± 850 BP probably corresponds to this ash layer. Similarly, a thin layer of tephra in the Sodmein Cave of the Red Sea Mountains, 1300 km south of Erciyes, has been linked to the Dikkartin Dağ eruption (Hamann et al., 2010).

These massive eruptions that deposited tephra over such large distances between 8000 and 9000 years ago probably had a profound effect on the Mediterranean cultures.

The geological features of the CAVP are further accentuated by the presence of active fault systems, such as the Central Anatolian Fault Zone (CAFZ). This fault system, characterized by left-lateral strike-slip faults and normal faults, plays a crucial role in shaping the topography of the region. The ongoing tectonic movements, evidenced by numerous earthquakes throughout history, underscore the dynamic nature of the Anatolian landscape (Fig. 2.6).

The Central Anatolian Fault Zone (CAFZ), with a total length of 450 km, is an active fault system and it is one of the most important neo-tectonic elements in Central Anatolia. Various field studies carried out by geologists show that the topography in the vicinity of this fault

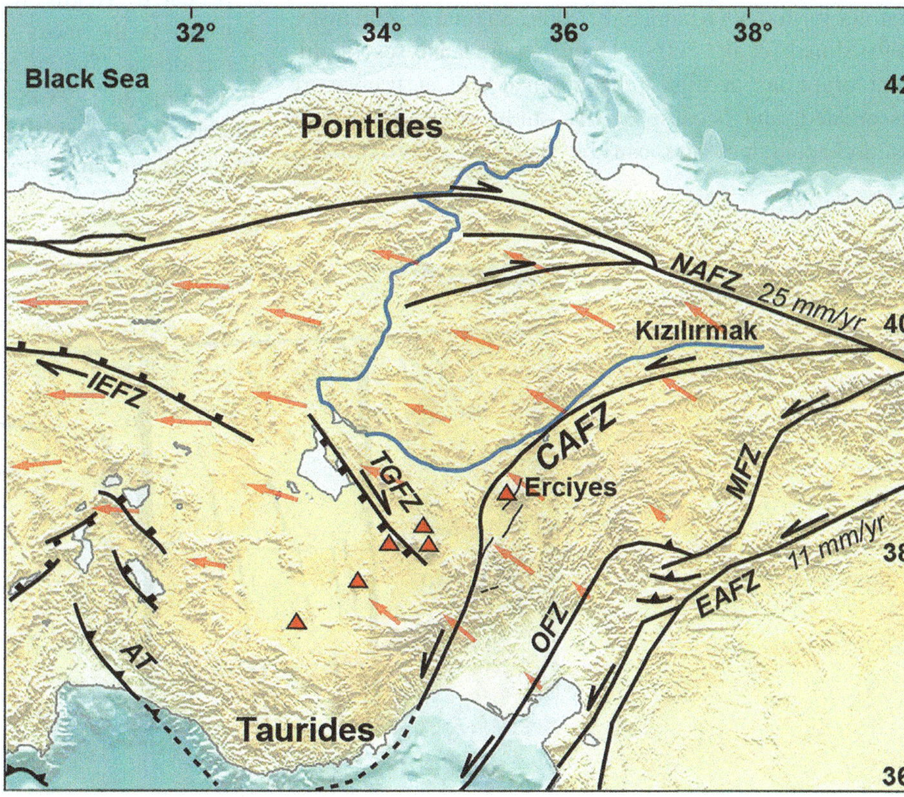

Fig. 2.6 Simplified active tectonic map of Central Anatolia. Red triangles indicate volcanoes, adapted from (Yıldırım et al., 2016) AT: Aksu Thrust, CAFZ: Central Anatolia Fault Zone, EAFZ: East Anatolia Fault Zone, IEFZ: İnönü-Eskişehir Fault Zone, MFZ: Malatya Fault Zone, NAFZ: North Anatolian Fault Zone, OFZ: Osmaniye Fault Zone, TGFZ: Tuz Gölü Fault Zone

is extremely young and has developed under the effect of active uplift, and that the tectonic activity causing this uplift is still active. The Erciyes region is one of the rare areas in Anatolia where active tectonism and volcanism are observed together (Kürçer et al., 2024; Tatar et al., 2000).

Moreover, it is quite natural that in a region of such great volcanism, there should be intense tectonic movement. There are several small active faults around the Ecemiş fault zone. Sultan Sazlığı, located in the middle of this fault zone, is a tectonic depression. This basin is surrounded by small but active faults such as Erkilet and Gesi in the north, Yeşilhisar in the southwest, and Develi in the southeast. This fault zone in the region is fully active and has produced numerous earthquakes throughout history (Ambraseys & Finkel, 1995).

In conclusion, the Central Anatolian Volcanic Province offers a fascinating glimpse into the geological history of Anatolia. From ancient volcanic eruptions to ongoing tectonic movements, this region continues to evolve, leaving behind a legacy of breathtaking landscapes and geological wonders.

2.3 Historical Background

The long history of human presence in Cappadocia goes back up to the Early Paleolithic Age. Kaletepe Deresi 3 Site, on the eastern slopes of Göllü Dağ in Niğde province, represents the longest Paleolithic sequence of Anatolia for today (Kuhn et al., 2009; Sagona & Zimansky, 2009; Slimak et al., 2008). Using tephra analyses of different eruptions at this archaeological site, a framework was created, and because of this study, some findings in this archaeological site can be dated to approximately 800,000 years ago. Researchers emphasize the importance of these findings as follows:

> … and provides a fundamental tephrostratigraphic framework to examine spatial and temporal variation in hominin behavior comparable to that of other regions, such as eastern Africa. (Tryon et al., 2009)

Apart from this site, findings dating to the Mid- and Late Paleolithic Ages were found at many different locations in the Cappadocia region. Çakmaktepe, Kömürcü and Kışla (Niğde), Suvermez (Nevşehir) are some of these settlements (Balkan-Atlı et al., 2007). Aşıklı Höyük (Aksaray), whose lowest layers date back to 7000 B.C., is the most important pre-Pottery Neolithic settlement of the Cappadocia region, with numerous dwellings excavated to date and many burial gifts found in intra-mural graves (Kuzucuoğlu et al., 2020; Özbaşaran & Duru, 2011) (Fig. 2.7).

Neolithic, Chalcolithic, and Bronze Age Sites are also well documented in the region. There are many more Neolithic sites in the region such as Köşk Höyük, Tepebağları (Niğde), Avladağ, Hasanlar, Sofular Höyük (Nevşehir). According to the data obtained as a result of archaeological excavations and surface explorations, almost all of the Neolithic Period settlements extended their settlement characteristics in the Chalcolithic Period (Esin, 1998). In some other settlements excavated, such as Çakıltepe Höyük, Yassıhöyük and Topakhöyük (Nevşehir), uninterrupted settlements are observed from the Early Bronze to the Medieval ages.

Located 16 km west of Kayseri, Kültepe (Kanis - Karum) is the most important Bronze Age settlement of Cappadocia. The archaeological excavations initiated by E. Chantre in 1893 in Kültepe, a settlement that is approximately 5.000 years old and the most important trade center in the region throughout its entire historical past, continued

Fig. 2.7 Aerial photo of Aşıklı Höyük (Google Earth Pro)

at different intervals and have been ongoing since 1948. Tens of thousands of clay tablets and numerous other findings unearthed during the excavations reveal the rich commercial relations between Mesopotamia and Anatolia between 1900 and 1700 BC, which began during the Early Bronze Age (ca. 3100 BC) when it was a rich city of Hatti. Kültepe is a settlement that provides valuable information about all phases of the Early Bronze Age, as well as the transition from the Early Bronze Age to the Middle Bronze Age south of the Halys River (Kulakoğlu, 2011; Kulakoğlu & Kangal, 2011) (Fig. 2.8).

In addition to these archaeological works, many different studies have examined the Assyrian, Hittite, Hellenistic, and Roman Empire periods of Cappadocia. There are Hittite inscriptions and monuments at many different locations in Kayseri and Nevşehir. Field surveys carried out in the region and excavations in ancient settlements such as Sobesos, Tyana, and Mokisos dating to the Hellenistic and Roman Imperial periods provide information about this period of the region (Akyürek, 1998; Darga, 1998; Equini Schneider, 1997; Tekin, 1998).

These sites' enduring value was conveyed by connecting roads that increasingly became stable lines between Mesopotamia and Anatolia; their descendant was the King's Road built by Darius I of Persia between Susa and Sardis.

A large part of the region defined by the great curve of the Halys/Kızılırmak and later to be known as Cappadocia began to be called "Tabal" after the fall of the Hittite Empire, at the start of the twelfth century BC; it extended west from Nevşehir and Nigde east to Gürün. At the south were the kingdoms of Tuwanuwa, near Niğde (its capital Tuwana or Tuwanuwa), Kybistra (today's Ereğli), and Melid (Melitene, now Malatya) to the east. Firsthand accounts of this region start to be more common in Assyrian documents of the eighth century BC, related to ties between the Assyrian Empire and Tabal. In these documents, Warpalawa is among the kings of Tabal, and his capital city, Tuwana or Tuwanuwa, is now recognized by most scholars as later Tyana, the current village of Kemerhisar. Its territory, named Tyanitis in classical times, perhaps coincided with the modern district of Nigde (ancient Nahita), a strategic area controlling the Cilician Gates, the most important pass from the Anatolian plateau through the Taurus Mountains to the Mediterranean (Sözen, 1998; Mora 2020).

The heartlands of Anatolia, including Cappadocia, were crossed by the armies that Alexander the Great led against the Persians, and suffered the consequences after his death. Among its governors were Ariarathes I, who proclaimed himself satrap of an independent Cappadocia in about 332 BC. After Seleucus Nicator defeated Lysimachus in 281 BC, his part of the region became "Seleucid Cappadocia," as distinguished from Pontic Cappadocia, which became the kingdom of Pontus.

Some scholars believe that the first true king of Cappadocia was Ariarathes III: none of the coins of his predecessors featured the word *king*, while all those minted by Ariarathes III did have that title. If those coins date to 255 BC, one may place the birth of the kingdom of Cappadocia in the mid-third century BC. His successor Ariarathes IV Eusebes married the daughter of Antiochus III, cementing the alliance between the Seleucids and Cappadocia. But after Antiochus' defeat in the battle of Magnesia in 190 BC, Cappadocia switched sides for the Romans and their ally, Eumenes II of Pergamon.

Fig. 2.8 Kültepe (Kanis-Karum) (photo A.E. Keskin)

During the reign of Ariarathes V Eusebes, both major cities of Cappadocia took his eponym: Mazaka as "Eusebeia by Mount Argaios," Tyana as "Eusebeia by the Taurus."

Following a subsequent period of Pontic domination, Bithynian incursions, and Cappadocian unrest, Rome intervened, ended the Ariarathid dynasty, and selected Ariobarzanes I, who was named Philoromaios, as the new king of Cappadocia in 96 BC (Erkiletlioğlu, 2019). There was substantial opposition, however, and Ariobarzanes I had a hard time keeping the throne despite Roman support from Sulla, Lucullus, and then Pompey, with intermittent depositions and restorations for almost three decades. Finally, in 63 BC, after Cappadocia had taken Kybistra (Ereğli) in Lykaonia and thus control of the road to the Cilician Gates, Ariobarzanes I, in the presence of Pompey himself, abdicated in favor of his son (Erkiletlioğlu, 2019). Ariobarzanes II Philopator, however, also had a troubled reign, and was killed in 51 BC. He was succeeded by his son Ariobarzanes III Eusebes Philoromaios, who managed to overcome political intrigues thanks to the help of Cicero, who had recently become governor of Cilicia.

In the civil war between Pompey and Caesar, Ariobarzanes III took Pompey's side, but after his defeat, Caesar forgave the king and incorporated Armenia Minor into Cappadocia, moving its borders eastward. Ariobarzanes III also aided Caesar in the war against Pharnaces II of Pontus, but in 42 BC he was killed by order of Cassius, one of the conspirators who had assassinated Caesar. Under the second triumvirate, Mark Antony supported Ariobarzanes' brother, Ariarathes X Eusebes Philadelphos, as successor, but then had him executed in 36 BC in favor of one Sisines from Komana, who had no connection with the family of Ariobarzanes and adopted the name Archelaos.

Despite Archelaos taking the side of Antony at Actium, Augustus not only took no action against him but in 20 BC allowed Cappadocia to keep Armenia Minor and added Cilicia, enlarging the kingdom to the east and south.

After a fifty-year reign, and despite his titles of Philopatris and Ktistes, Archelaos was summoned to Rome by Augustus' successor Tiberius, and died there. Shortly after his death, Tiberius' nephew and adopted son Germanicus defeated the king of Armenia and made Cappadocia a province in 17 AD (Erkiletlioğlu, 2019) (Fig. 2.9).

At a time when Armenia and Mesopotamia were not yet Roman, the location of the new province again made it a borderland whose military and political control was of great strategic importance as the eastern wall of the Roman Empire (Sözen, 1998).

Cappadocia's strategic importance was again attested when Vespasian added legions in Cappadocia to face the continuous raids of the barbarians and appointed a consular legate instead of an equestrian. In 70 AD, Vespasian joined Armenia Minor to Cappadocia and made the combined province a frontier fortification. It remained, under various provincial redistributions, part of the Eastern Empire for centuries (Mitchell, 2018).

This union of provinces was not stable, however, and various separations and recompositions occurred over time. Still, during the emperor Trajan's reign (98–117 AD), Cappadocia continued to function as a buffer region between the Roman world and the threat of Parthia, and even more so when Hadrian rolled the eastern border back to the Euphrates (Fig. 2.10).

Cappadocia again took on a crucial strategic and logistic role between the end of the second and mid-third century, with Caesarea and Tyana as important military areas: first in 193–194 AD, during Septimius Severus' war against Pescennius Niger, the governor of Syria who had made the legions acclaim him as emperor; then due to the repeated Persian campaigns under Alexander Severus, Gordian III, and other third-century emperors; and especially in 260 AD, when the emperor Valerian I was captured at Edessa (modern Urfa) by the Persians.

We would like to emphasize one point here: The region defined as Cappadocia today is a very small part of ancient

Fig. 2.9 Coin of Tiberius minted in Caesarea—Cappadocia AD 14–37. Silver drachm with Mount Argaeus on the back side (www.cngcoins.com)

Fig. 2.10 Coin of Hadrian minted in Caesarea—Cappadocia AD 117–138. Silver didrachm with Mount Argaeus on the back side (www.cngcoins.com)

Cappadocia. This region was a province of the Roman and the Byzantine empires. Significant changes were made in the formation of this province during both empires. As stated by various scholars, such as W. M. Ramsay, in AD 314 Cappadocia was the largest province of the Roman Empire, covering an area of at least 50,000 km^2, almost all of Aksaray, Nevşehir and Kayseri, and even part of the modern provinces of Kırşehir, Yozgat and Niğde (Ramsay, 1890) (Fig. 2.11).

Despite this fact known, the touristic and scholarly attraction of the rock-cut churches full of frescoes located in the Nevşehir–Ürgüp–Göreme triangle has led to the mistake of the acceptance of this very small area as the entire "Cappadocia."

Between the later third century and the beginning of the fourth, the region was also attacked by Goths and the Palmyrene Empire, and under Diocletian Cappadocia was divided into two sectors: the larger, western one retained the name of Cappadocia, while the smaller eastern part, including the cities of Komana and Melitene, was merged with Armenia Minor. In 371 AD, Valens further subdivided the aforementioned region of Cappadocia into Cappadocia Prima and Cappadocia Secunda, with their capital cities Caesarea and Tyana, respectively (Mitchell, 2018).

The fourth century was a period of prosperity, when Cappadocia produced three bishops and Fathers of the Church, Gregory of Nazianzos, Gregory of Nyssa, and Basil of Caesarea. Their missionary work was made possible by the fact that their homeland was a crossroads of the Anatolian peninsula. But from the fifth century on, the Byzantine emperors had to build a line of fortifications along their eastern border in Cappadocia (including Caesarea), to protect their empire against the Persians, and later, the Arabs (Rosada et al., 2024).

The duration of tranquility following the definitive cessation of Arab assaults and incursions was, however, exceedingly brief. A new threat soon emerged: the expansion of Turkic tribes from the eastern boundaries into Anatolia. The initial Seljuk invasions commenced circa 1066 (Hild & Restle, 1981).

Consequently, we cannot dismiss the possibility that Cappadocian shelters were never abandoned, but rather that the resident people maintained or even extended the subterranean defenses. It is conceivable that the inhabitants believed the menace of Turkish incursions would be mitigated through transient and intermittent measures, similar to those employed against the Arabs, which the Byzantine imperial authority would curtail, failing to anticipate that this occasion would lead to a lasting occupation of the territory, rendering the shelters ineffective:

> Cappadocia passed with hardly any struggle into the hands of the Seljuk Turks immediately after 1070. (Ramsay, 1897)

In this example, the completion of at least some shelters may have occurred later, specifically between the mid- and late eleventh century, as posited by Veronica Kalas (2019) regarding the most possible occurrences in the case of Selime. Subsequent to that date, the territory never

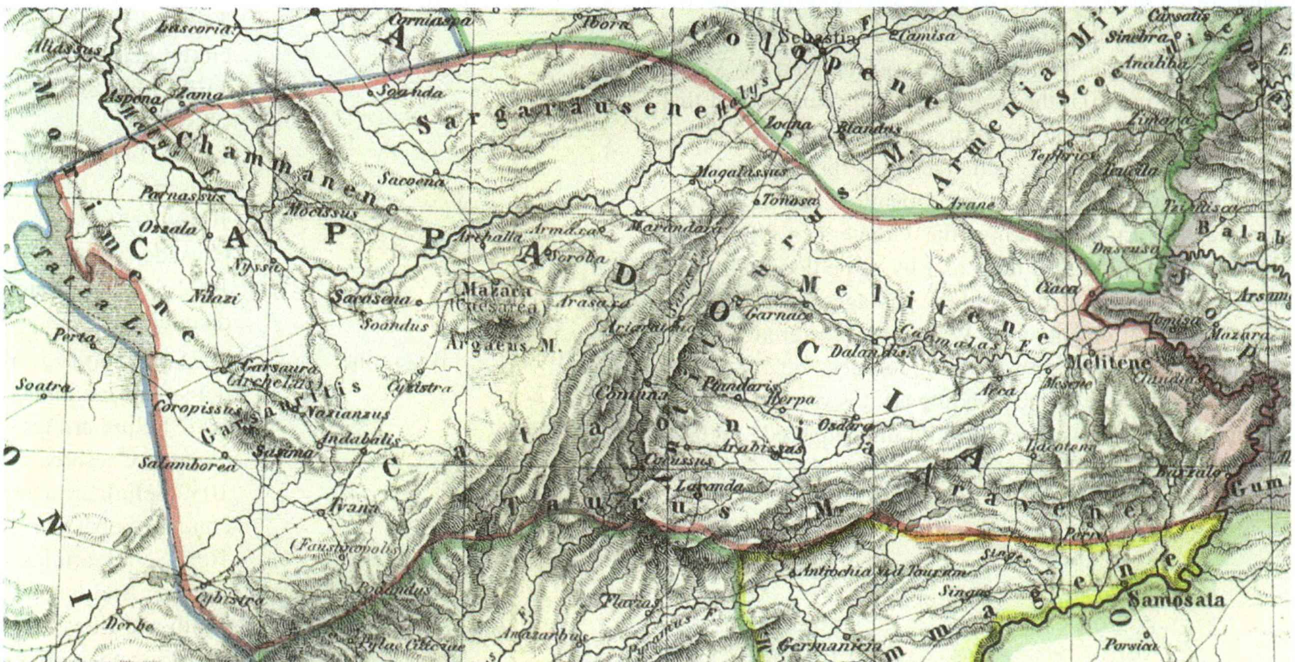

Fig. 2.11 The province of antique Cappadocia in Kiepert's map (Kiepert, 1903)

returned to Byzantine control, and theoretically, the shelters should have been obsolete. Nevertheless, it is imperative to acknowledge that subsequent periods were undoubtedly tumultuous.

For example, Kyzistra's territory (today Yeşilhisar), which was home for a large rock-cut settlement with a monastery, was contested between Byzantines and Arabs starting from the second half of the seventh century, then burned in 912. After the mid-eleventh century, Gagik II of Ani, the Armenian king, waged war against the Byzantines. He was apprehended and executed in Kyzistra in 1079. During this period, as previously mentioned, the Seljuks arrived and established themselves in the area, initially in western Cappadocia, and then conquering the entire region and beyond.

Subsequent events likely prompted the inhabitants to utilize the underground shelters once more, particularly during the Mongol conquests of Kayseri and Niğde in 1243 and 1257, which expanded their dominion over all of Cappadocia until the mid-fourteenth century.

Turbulence persists during the conflicts between the Ottomans and local rulers, exemplified by the capture of Nevşehir from the Karamanids in 1466. From 1473, the territory permanently remained under Ottoman control (Aldehuelo, 2003). Therefore, even just considering this small information about a permanently troubled and uncertain situation, it would be quite difficult to find a period when shelters had not been essential for the inhabitants' security.

The last document, about the use of an underground shelter for the defense, refers to a relatively recent episode, occurred near Misli, today Konaklı, a locality about 25 km northwest of Niğde. In this village, as reported by White after his visit in 1903, the population still remembered that the inhabitants successfully retired "under the ground" to escape the advancing of Egyptian army of Ibrahim Pasha:

"When Ibrahim Pasha invaded Turkey half a century ago, with the Egyptian army, the villagers of Misli fled below ground, cutting off their rear by stone doors like mill stones, which they rolled across the passageways. The army could not force an entrance. When they lowered buckets into the wells to draw up water, the refugees below cut off the buckets, and finally the invading army swept on, leaving a village of cavate dwellers behind it unconquered" (White, 1904).

2.3.1 Historical Timeline for Cappadocia

Date AD	Historical Periods in Cappadocia	Events
1923–present	Turkish Republic	1923 Population exchange between Greece and Turkey, est. 60,000 Cappadocian Greeks migrate to Greece
~1450–1923	Ottoman Empire	1832 Ibrahim Pasha of Egypt defeated Ottomans and occupied Cappadocia Seventeenth century agricultural crisis Late fifteenth and sixteenth centuries generally prosperous
1308–1450	Interregnum	1402 Timur's invasion of Anatolia Fourteenth-century Post-Seljuk instability, local dynasties or short-lived occupations
1080–1308	Seljuks	1308 Anatolian Beyliks 1243 Mongol invasion Twelfth-century agricultural crisis 1100–1308 Sultanate of Rum 1147–1149 Second Crusade 1095–1099 First Crusade 1092 Seljuk Empire disintegrate
950–1080	Byzantine	1071 Byzantine army defeat at Mantzikert 1067 Seljuk armies conquer Caesarea 1059 Seljuk armies conquer Sebasteia 1057 Seljuk armies conquer Melitene Tenth and Eleventh centuries peak of Christian mural art

Date AD	Historical Periods in Cappadocia	Events
~628–950	Dark Ages	950 Strengthening of the Byzantine Empire 934 Arabs conquer Melitene 730–787, 815–843 Iconoclast periods 781–833 Arab raids 709 Arabs conquer Tyana 647 Arabs conquer Caesarea Seventh-century Cappadocia becomes military frontier zone between Byzantine and Arabs
~330–628	Byzantine	604–628 Sasani wars 572 Sasanians conquer Melitene 540–590 Repeated outbreaks 540–542 Justinian Plague Fifth–sixth centuries earliest surviving rock-cut churches, major church building 395 Roman Empire split into east and west 371 Valens divided Cappadocia into Cappadocia Prima and Cappadocia Secunda “Cappadocian Fathers” Basil the Great (330–379), Gregory of Nyssa (c. 335–c. 395), Gregory of Nazianzus (329–389)

2.4 Rock-Cut Architecture

Artificial cavities and rock dwellings are common not only in the Cappadocia region but also in almost every region of Anatolia. Although located hundreds of kilometers apart and spanning thousands of years, all these structures, excavated in different rocks, such as volcanic and limestone, serve similar but varied purposes from residential to the storehouse, and from religious to defensive structures.

Singular structures, especially rock-cut tombs, exist all over Anatolia and have been a common tradition throughout Anatolia history from west to east. Such tombs excavated in rock walls or underground have a history of thousands of years and include Lycian, Greek, and Roman tombs in Western Anatolia as well as Urartian tombs, dated to 800–700 BC, which are common in the eastern provinces (Çevik, 2000). Roman rock-cut tombs with ornamented front façades carved into a flat wall are found in or around many different antique sites such as Telmessos, Kaunos, Myra in Western Anatolia. There are also Roman rock-cut tombs carved directly into the bedrock on the ground, with several klines and occasional sarcophagi, examples of which can be seen throughout different provinces of southeast Anatolia, such as in Diyarbakır-Hilar or Gaziantep-Dülük.

In addition to the thousands of rock-cut tombs found in numerous graveyards throughout Anatolia, several singular rock-cut structures have been carved in provinces such as Konya—Kilistra and Sille, Bitlis, Ankara-Nallıhan, Batman, Mardin—Midyat, or Çorum, when the geological formation is suitable for carving. Although some of these singular structures are churches, as in the example of Konya—Sille (Yamaç & Gilli, 2018), rock-dwelled houses and defense structures are also encountered locally (Bixio et al., 2011; De Pascale & Bixio, 2010).

On the other hand, unlike these singular structures, there are also dense rock dwellings carved close to one another in different parts of Anatolia. Apart from Cappadocia, one of the most common places where such structures are seen is the region known as the “Phrygian Highlands,” in the provinces of Eskişehir—Afyon in Northwest Anatolia. There are hundreds of rock-cut shrines and settlements in this area of approximately 4000 km^2. The rock-cut structures in this region were first carved by the Phrygians in 600–700 BC; some of the structures were used for residential purposes or converted into churches over time. In addition, many more rock-cut churches were carved during the Byzantine period, and Uçkan describes in detail 42 rock-cut churches located in this region that can be dated to this period (Uçkan, 2010). It is very difficult to determine which of the structures outside the protected areas were carved during the Phrygian period because of the constant changes to the structures over time; there is no clear inventory of such structures. Aside from the difficulty of collecting all these rock-cut monuments, shrines, cisterns, and dwellings, found in a very large area and carved over at least 2700 years, in a single inventory, it is debatable how accurate such a study would be. Nevertheless, the studies of Berndt-Ersöz and Haspels on the rock-cut structures found in this region and dated to the Phrygian period are quite comprehensive (Berndt-Ersöz, 2006; Haspels, 1971). In "Phrygien und Pisidien," written by Belke and Mersich and published as the seventh book of the "Tabula Imperii Byzantini" series, there is both an inventory of Byzantine buildings in this region and a comprehensive bibliography of works about this period in the region (Belke & Mersich, 1990).

In Karaman Province in Central Anatolia, there are also numerous rock-cut structures carved into the neritic soft limestone. Although first discovered by the Russian geographer Pierre De Tchihatcheff in the middle of the nineteenth century, only some of these structures which are concentrated around Manazan and Gödet villages and which have become a large cliff settlement have been subjected to scientific studies. In the second volume of his monumental work "Asie Mineure: Description physique, statistique et archeologique de cette contree," published in five volumes between 1853 and 1860, Tchihatcheff stated regarding Gödet Village, "This rock settlement, on the walls of a deep canyon, is very impressive" (De Tchihatcheff, 1853–1860). These cliff settlements at Gödet Village, which we studied in 2012, were composed of an enormous number of dwellings on both walls of an 11 km long canyon, and some of these structures were still in use. This troglodyte settlement in the canyon is very large and requires extensive research (Yamaç, 2012).

Soft limestone found in Southeast Anatolia, especially around Hasankeyf, Urfa, and Gaziantep, has been carved for centuries and many different structures, from houses to churches, have been built using it. There are hundreds of dwellings carved on the walls of the Tigris River and ancient citadels in Hasankeyf, and in Urfa and Gaziantep, both on the walls of the Euphrates River and inside the cities. Among these structures, some of which have been studied by us in recent years, there are houses, churches, and even a five-story monastery. Although a rock-cut church with Armenian inscriptions on its walls that we found during our studies around Urfa—Halfeti, along with the well-known Hromgla Castle on the opposite shore, can be dated to the twelfth or thirteenth century, we are aware that this date cannot be valid in any way for other dwellings in the region. Another interesting fact is that some of the dwellings in these regions are still being used by local people. Although some of the cave dwellings in the city of Gaziantep and Urfa are used for storage purposes, many artificial caves in both cities are now restaurants or cafés. Apart from two different inventories and some articles, hundreds of rock-cut structures in this region have not been the subject of scientific research until now (OBRUK, 2012, 2019; Yamaç, 2013; Yamaç & Okuducu, 2017).

In contrast, in eastern Anatolia, especially in Ani and Ahlat, hundreds of rock-dwelled settlements were explored, surveyed, and documented during years of archaeological work. Dwellings carved in volcanic rocks in the east of Kars, on the slopes of the ancient Armenian site of Ani facing the Arpaçay River (Ahurian) and in the citadel attracted the attention of various researchers in the early 1900s. In 1915 those structures were investigated by Kipshidze. After his death, Kipshidze's notes were published by Tokarski in the Armenian Academy of Sciences in 1972. These notes mention the impressive number of 823 underground structures, mostly dwellings, but also churches, passages, cemeteries, and dovecotes in Ani (Kipshidze, 1972). Most of these artificial cavities found in Ani were extensively researched and published by Centro Studi Sotterranei of Genoa (Bixio et al., 2009).

Although these dwellings in Ani can be dated to the ninth–tenth centuries AD, like many other structures of the city, hundreds of years of uninterrupted use make it impossible to determine the first construction date. Although a dense rock settlement is observed in this ancient site, researchers working here do not accept "A huge Ani underground," proposed by previous researchers who claimed that more than a thousand underground structures, where almost all the inhabitants of Ani lived in underground dwellings until the structures on the surface were built (Bixio et al., 2009).

During the four-year surface investigations in Ahlat, on the west coast of Van Lake, more than 400 rock-cut structures were identified and surveyed in 17 different zones. Although it is difficult to determine the age of all these structures carved in the thick pyroclastic rocks piled up as a result of intense volcanism in the region, which is not unusual in the case of artificial cavities, the structures are likely to date to a period between the twelfth and fourteenth centuries, both by comparing the architectural features of the Armenian church and by examining the potteries found during the underground tunnels excavations (Bixio & De Pascale, 2011; Bixio et al., 2013, 2014, 2015).

Few years ago, a comprehensive survey of cliff settlements and rock-cut churches was started at 26 different locations in northeastern Anatolia, in the region stretching from Ardahan to Georgia, especially in the Kura River basin. During the first season of this archaeological survey a cliff settlement with a total of 68 different rock-dwelled structures, containing three rock-cut churches, was explored and surveyed (Arda, 2017) (Fig. 2.12).

However, in terms of rock-cut structures, the Cappadocia region is very different from all these other regions we mentioned above. The difference is that besides the sheer number of rock-cut churches or dwellings in this region, there is a troglodyte civilization in a very large area that lasted hundreds, perhaps even thousands of years.

Despite all these studies on the history of Cappadocia, very limited archaeological or historical data are available about the underground settlements in this region. When

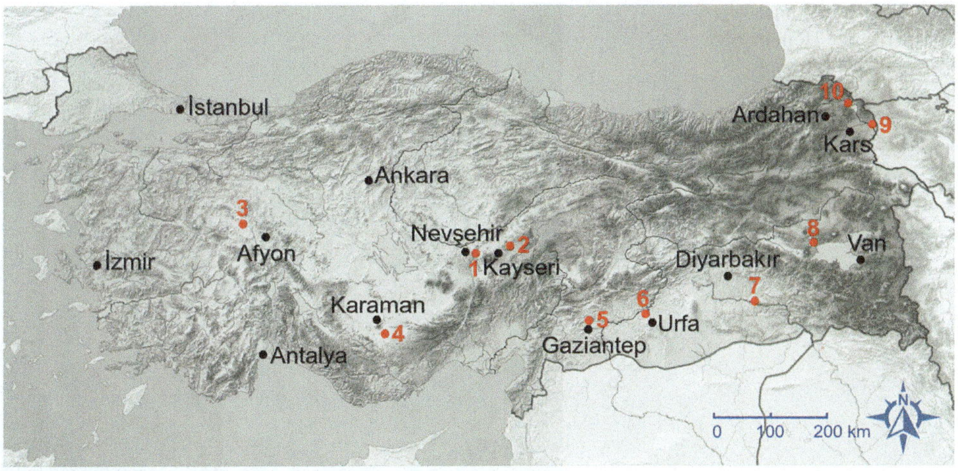

Fig. 2.12 Locations of some dense cave settlement areas mentioned in this chapter. Signs do not indicate a specific point, but a larger area in that region: 1 Cappadocia, 2 Koramaz Valley, 3 Phrygian Highlands, 4 Gödet Village, 5 Gaziantep, 6 Halfeti, 7 Hasankeyf, 8 Ahlat, 9 Ani, 10 Kura River (Compiled by A. Yamaç, Google maps)

the volcanic rocks covering the entire area first started to be dug, as well as when the inhabitants started to use the underground as their dwellings, is unknown. It is possible that local people dug this easy-to-carve soft rock for centuries, and that a troglodyte civilization was established in the region that is known as Cappadocia today.

Although there is a general tendency to date this troglodyte civilization in Cappadocia to the Byzantine Empire period because of rock-cut churches and underground defensive structures thought to be dug by local people to protect themselves from Arab raids from the seventh to eleventh centuries, Kostof and many other researchers think, “there is no reason to disbelieve that the practice was more ancient than Christianity” (Kostof, 1972).

The earliest dated rock-hewn structures encountered in Cappadocia are probably Roman rock-cut tombs and these numerous Roman rock-cut tombs seen in the region today confirm Kostof’s proposal. There are several rock-cut tombs in Göreme, Mazıköy (Ousterhout, 2017), and south of Kayseri, especially the vicinity of Ayşepınar and Yeniköy (Durukan, 2012) (Fig. 2.13). On the other hand, numerous other Roman rock-cut tombs close to the settlements have been changed over time for different purposes.

The continuous use of the rock-hewn structures for thousands of years has made precise dating nearly impossible. As many scientists researching the region have argued, it is almost impossible scientifically to determine the date of the first carving of a rock settlement. In time, a Roman rock-cut tomb becomes a house, then a warehouse, or a barn. The walls are destroyed or built and you cannot see many of these changes because not much remains of the past structures. Perhaps a rock decoration indicates that a warehouse used to be a Roman rock-cut tomb or a rock-cut structure that used to be a church and which later became a dovecote.

The persistent utilization of rock-cut spaces introduces an additional challenge, given their capacity to maintain their fundamental integrity over time. Numerous carved spaces have undergone conversion to new functions and have been continuously employed beyond their initial context of creation. These spaces are readily adapted to new uses and transformed, both in reality and perception. A significant number of carved spaces have been subsequently repurposed as sheepfolds, stables, dovecotes, and other functions. However, this adaptation often results in the loss of original features, and the subsequent carvings are often challenging to identify with precision, except when they manifest a discernible irregularity (Günay, 2024).

At the same time, for building a defensive structure, an artificial underground cavity is always preferred over a surface structure. These subterranean defense systems are widespread in Anatolia and can be found in hundreds of locations worldwide, from northern France to Vietnam. It is known that the Cappadocia region contains countless defense constructions known as “underground cities” or “underground shelters,” each of whose tunnels is guarded by a millstone door. Defensive structures in the form of an artificial cavity are not the only examples of the so-called “underground cities” whose tunnels are protected by millstone doors and which weave the bottom of a village like a cobweb. “Rocky castle villages,” examples of which are seen in Uçhisar and Ortahisar in Cappadocia, are also excellent defensive structures.

It is often claimed that underground structures were more likely to have been temporary settlements, for use primarily in times of insecurity. Millstone doors are found frequently in underground shelters in Cappadocia. The difficulty in moving them, and the fact that they are usually cut out of harder rock than the cavity itself, indicate that these were not meant for everyday use, but only in cases of emergency, when they would be rolled into place to block the entrance to the refuge. The discovery of spaces that may have been small chapels within the underground settlements, and the fact that all artifacts found in the underground spaces date back to the fifth to tenth centuries,

Fig. 2.13 Ayşepınar—Kayseri, Late Hellenistic rock-cut tombs (photo A. Yamaç)

suggest main occupation during the Byzantine period. Today, the underground structures can be accessed from individual houses in the villages on the plateau above, and this direct connection between permanent villages and temporary refuges would have made sense particularly during the seventh to eleventh century, which was a period marked by seasonal Arab attacks. Moreover, differences in the quality of carving suggest that the underground settlements must have been enlarged continuously by successive generations.

A classification of settlement types based on the differences between *secular* and *religious* function is not an easy task in Cappadocia. While it cannot be denied that the region was home to different forms of monasticism to some extent, several complexes that were initially identified as monasteries are now considered to be secular establishments. On the other hand, simple villages, carved either as cliff or fortress settlements at one extreme, and isolated or groups of elite courtyard houses, adorned with carved facades, at the other, are plentiful in Cappadocia. To make it more complicated, the eremitic way of life was not so far removed from the daily life of the population, and obviously, the peculiar landscape of Cappadocia, while offering various means of escape, also presented various opportunities to exhibit one's status.

To date, in the absence of inscriptions making precise dating possible, religious wall paintings and architecture have been the most reliable physical evidence for dating in Cappadocia, although there is a question of the contemporaneity of carving and painting. The earliest examples of churches date back to the sixth or seventh century and are few in number; while the majority are believed to fall between the ninth and eleventh centuries, prior to the arrival of the Turks. The only plan type that pinpoints with a sufficient degree of reliance a closer date interval is the cross-in-square plan, which was a distinctively middle Byzantine innovation. On the other hand, a comparative study of carving techniques and carved decoration can be of aid in proposing at least a relative chronology. In this respect, churches bearing common features such as large apses framed by precisely semicircular arches, large cross reliefs on walls and ceilings, stepped cornices with dentils, medallions, and reliefs of palms are identified as early Byzantine. Interestingly, precisely shaped geometry, sharp corners, and fine and regular marks of pickaxes are also common qualities of early churches, although this alone does not exclude a later date. The scarcity of archaeological evidence from the mid-seventh to the ninth centuries—which was a period of insecurity when Arab attacks were commonplace—suggests a rupture of architectural productivity, which in turn may be an indication of a decrease in population (Öztürk, 2017).

The exact number of rock-cut structures in Cappadocia is unknown, and no comprehensive inventory is available. Apart from rock-cut structures, even the total number of rock-cut churches, on which so much research and study have been done, is debatable.

2.5 Previous Researches in Cappadocia

The different natural formations of Cappadocia and numerous rock-cut structures dug in these formations, especially the rock-cut Byzantine churches, have drawn the attention of western travelers since the eighteenth century and have been the subject of various scientific studies. These works and studies started for the frescoes in numerous rock-cut churches found in the area have expanded to other rock-cut structures.

Today, the articles and books on the rock-cut structures—from the underground shelters to the hydraulic structures and from the dovecotes even to the rock-cut apiaries—of the Cappadocia number in the thousands. A comprehensive anthology of all the books and articles written about Cappadocia today would likely constitute a substantial volume in itself. Published as the second book of Tabula Imperii Byzantini series in 1981, Hild and Restle's "Kappadokien" is highly comprehensive in terms of bibliography as well as including a detailed inventory of a wide area (Hild & Restle, 1981). The works of Restle and Jolivet-Levy on rock-cut churches and frescoes also have a long bibliography (Jolivet-Lévy, 2015; Restle, 1967).

Two works contain comprehensive bibliographies on all other rock-cut structures in Cappadocia except churches are: Bixio, A., Bixio, R., & De Pascale, A. "Göreme and Şahinefendi" and Bixio, R. "Cappadocia: Records of the Underground Sites" (Bixio, 2012; Bixio et al., 2023).

The first author to mention Cappadocia is Herodotus (c. 484–c. 425 BC). In his book "The Histories," he mentions that in 547 BC, Thales of Miletus changed the directional flow of the Halys River, allowing the armies of Croesus to cross the river and thus to take Pteria, the capital of Cyrus in northern Cappadocia (Gilli et al., 2014; Herodotus, 1920). Following Herodotus, many ancient writers provided some general information about the region over the centuries but did not offer much detail about it. For instance, it is strange that although Pliny and Strabo mentioned Cappadocia in their works written in the first century AD, there is no information about the most important feature of this region, namely the troglodyte life, which we almost certainly know existed at that time. Almost all works related to this area written during the Byzantine period are of a religious basis; therefore, in the ancient documents we have today, there is almost no information about Cappadocian cave dwellings. The only exceptions are the short paragraphs mentioning these structures in three different ancient sources.

Paul Lucas was the first western traveler who saw Cappadocia. While passing through Asia Minor, Lucas, who was commissioned by the French King Louis XIV in 1704 to explore the eastern countries, saw near the Halys River (Kızılırmak) a large number of "pyramidal houses" with conspicuous entrances and stairs and big windows to light all the rooms and paintings inside; likening the fairy chimneys as "monks with hoods" and the rocks on the fairy chimneys to the busts of "Mother Mary holding baby Jesus." Since then, these now-famous fairy chimneys have greatly excited the imaginations of travelers. Over time, the hard layer of limestone on Cappadocia's plateau has worn away, exposing the soft volcanic tuff underneath. Winds and rains then shaped this tuff into cone-like formations, often leaving a limestone block at the top of the cone. With these limestone blocks imagined as monks and virgins, Lucas created a spiritual landscape for the region. Also, he thought at first that the curious houses might be the ancient dwellings of Christian hermits (Lucas, 1712) (Fig. 2.14).

Fig. 2.14 This engraving by Lucas is the first visual depiction of Cappadocia (Lucas, 1712)

It would take more than a century for a more detailed description of Cappadocia after Lucas to emerge. Charles Texier revisited the region in 1833 and then in 1837. Countless engravings and plans in Charles Texier's three-volume comprehensive work (Texier, 1839) revealed the rock-cut structures of Cappadocia for the first time and provided a more realistic description of Ürgüp and Göreme (Fig. 2.15).

English traveler W. F. Ainsworth, who visited Cappadocia during 1840, wrote:

> Turning up a glen which led from the river inland, we found ourselves suddenly lost in a forest of cones and pillars of rock, that rose around us in interminable confusion, like the ruins of some great and ancient city. (Ainsworth, 1840)

A few years after these two travelers, W. J. Hamilton, who passed through Cappadocia as part of a long Anatolian crossing, writes that he stayed in Tatlar Village and examined some of the frescoed rock-cut churches in the vicinity. Again, as he states in his book, he studied the churches in Ürgüp in detail, as he was influenced by Texier's notes, and from there he went to Uçhisar (Hamilton, 1842) (Fig. 2.16).

Pierre de Tchihatcheff carried out six different trips and research in Anatolia between 1848 and 1853, as a result of his studies, he became the first scientist to evaluate the physical geography, geology, and botany of Cappadocia with a scientific approach. During his research in the region, he discovered numerous fossils and endemic plants; in addition, he carried out the first investigations on volcanic sedimentation and geological stratification (De Tchihatcheff, 1853–1860).

Although not as comprehensive as Texier and Hamilton, Henry Tozer came to the region in 1881. His memoir describing Kayseri and Nevşehir is interesting in several respects. Tozer met with Christian and Muslim villagers living in the region, and in addition to his extensive ethnographic studies, he made a very difficult climb to the summit of 3916 m high Erciyes Mountain during his stay in Kayseri (Tozer, 1881).

Fig. 2.15 Facade of a rock-cut structure in an engraving by Texier. This structure, which does not exist today, is thought to be the courtyard façade of the Bezirhane in Avcılar, near Ürgüp (Kalas, 2004; Texier, 1839)

From the late 1800s to the early 1900s, Western travelers started to do more detailed and scientific research in Cappadocia. They were amazed by the region's unique landscape and the fact that the local people spoke Greek. This was surprising because the Islamic conquests had supposedly separated Cappadocia from the rest of the Hellenic world. Scholars were very interested in the oral traditions of Cappadocia. This interest, combined with the development of Modern Greek nationalism, led to an effort to prove that the Greeks of the region were directly related to the Greeks of late antiquity. The Christian and Greek monuments of Cappadocia helped strengthen this effort throughout the Hellenic world in the late nineteenth century. This effort was to define a Greek national, ethnic, and religious identity. At the time when these political views were being supported by the features of architectural structures, Josef Strzygowski published his first polemical work, Orient oder Rom. This work and Kleinasien, published two years later, examined the origins of Christian monuments built in Anatolia in the Middle Ages. The second work, which covered Cappadocia's architecture in detail, basically argued that this architectural style was not Hellenic but Oriental (Strzygowski, 1901, 1903).

After Strzygowski's Kleinasien, Gertrude Bell and Sir William Ramsay produced the next comprehensive study of the early Christian churches of Asia Minor. On the other hand, of all the early architectural historians to study Asia Minor, Hans Rott produced the most comprehensive work. Rott had accompanied Strzygowski on many expeditions and was the first to methodically attend to inscriptions ignored

Fig. 2.16 An engraving of Uçhisar—Cappadocia (Hamilton, 1842)

by Bell. In his work, Rott covered both the early Christian and Byzantine periods and he was the first to present buildings from the Seljuk period as well. His broad treatment of Cappadocia included a large number of newly discovered rock-cut monuments (Ramsay & Bell, 1909; Rott, 1908).

The works of these researchers constitute an example of successful scientific studies. These works, which mainly depict the inscriptions, rock-cut churches, and frescoes in this region, can be considered as a "first step" even though they are insufficient to cover the whole region.

One of the earliest archaeologists to record the prehistoric sites of central Anatolia, especially those of Cappadocia, was Ernest Chantre. During his research in a wide area for two years, he discovered and published many inscriptions as well as prehistoric findings (Chantre, 1898).

J.R.S. Sterrett's publications about his trips in Anatolia and especially his article "The Cone Dwellers of Asia Minor" published together with 52 photographs of J.H. Haynes in National Geographic Magazine in April 1919 are remarkable. To Sterrett's astonishment, he found people still living in rock-cut dwellings when he visited Cappadocia and wrote that (Fig. 2.17);

> [t]here is no earthly reason why they should live there, as the country is safe and land abundant.

Although the publication date of this article is 1919, Haynes' photographs are from the years 1884–1887 and are probably the oldest known photographs of Cappadocia Guillaume de Jerphanion, born at Pontevès in 1877, was a French Jesuit, archaeologist, and explorer of Cappadocia.

The magnificent four-volume work of Jerphanion, "Une nouvelle province de l'art byzantin, les églises rupestres de Cappadoce" (Jerphanion, 1925–1942) is perhaps the most fundamental work on the rock-cut churches of Cappadocia. This sizeable catalogue presented a modern study of rock-cut church paintings from the valleys of Göreme and Soğanlı in eastern Cappadocia (Fig. 2.18).

It was written following three trips to Cappadocia, the first in August 1907, the second and the most important

Fig. 2.18 A fresco from the Kılıçlar Church, Göreme (Jerphanion, 1925–1942) (drawing Ernest Mamboury 1912)

Fig. 2.17 One of the earliest photographs of Uçhisar—Cappadocia by J.H. Haynes dated 1884 (Ousterhout, 2011)

from August 1911 to January 1912, and finally the third in August–September 1912.

In the four volumes that constitute the complete work, a whole series of monuments from different periods between the ninth and the thirteenth centuries, and mainly located in the Ürgüp region, are listed in a methodical and homogeneous manner. The complete work therefore can be considered as the database of the period, and after so many years, still continues to be a reference. In the first volume of this work of Jerphanion, there is also a comprehensive list showing the settlements visited by all the western travelers who researched the region before the 1900s.

In contrast to the extensive exploration carried out since the 1800s on this small piece of Cappadocia located in the Nevşehir–Ürgüp–Avanos triangle, other parts of Cappadocia have been little explored.

As Ousterhout observes:

> The high plateau of rocky Cappadocia, bounded by Kayseri in the east, Aksaray (Koloneia) in the west, and Niğde (Tyana) in the south, has been of considerable interest to scholars and tourists alike because of the rock-cut settlements from the Byzantine period. (Ousterhout, 2017).

Since the 1950s some other churches that are not included in Jerphanion's work have been researched and published in different books and articles. For instance, rock-cut churches located in Peristrema Valley (Ihlara) were studied by Nicole and Michel Thierry in 1963 (Thierry & Thierry, 1963).

The number of publications about Cappadocia increased as a result of the post-1960s spike in scholarly interest in the area. For a number of factors related to its physical terrain, geology, geopolitical significance, and artistic and architectural legacy, it attracted a lot of interest from Byzantine art and architecture academics and archaeologists. Roman and especially Byzantine constructions have a remarkably high preservation rate due to the widespread integration of building into the living rock; this fact has naturally drawn attention (Haldon, 2024).

Among the academics who conducted research on Cappadocia during this period, the works of Marcel Restle, Frederich Hild, Luciano Giovannini, Catherine Jolivet-Levy, and Robert G. Ousterhout are particularly noteworthy.

Although there are many different bibliographical databases on Cappadocia, *"Byzantine Studies in Turkey: A Bibliography,"* edited by Şahin Kılıç and Buket Kitapçı Bayrı, is the most comprehensive work written on the subject to date (Kılıç & Bayrı, 2021).

Nevertheless, apart from a few articles, there has not been much scientific work on other rock-cut structures except the churches of the area. Most of the works describing the cliff settlements or structures in the valleys are at the "touristic guide book" level. Apart from churches, the most comprehensive research and publications on rock-cut or underground structures in the region are about underground cities. The first study on this subject belongs to Urban, followed by works of Yörükoğlu and Triolets (Triolet & Triolet, 1993; Urban, 1973a, 1973b, 1973c; Yörükoğlu et al., 1989). Initiated in the early 1990s by the National Commission for Artificial Cavities of the Italian Speleological Society in cooperation with the Turkish Ministry of Culture and the Archaeological Museum of Nevşehir, the primary objective of this project was to conduct a comprehensive survey of the area and undertake a systematic exploration of the rocky and underground sites. After years of work, numerous articles and books have been published, showing the results of extensive studies (Bertucci et al., 1995; Bixio, 2012; Bixio et al., 2002).

References

Ainsworth, W. (1840). Journey from Angora by Kaisariyah, Malatiyah, and Gergen Kal'ehsi, to Bir or Birehjik. *The Journal of the Royal Geographical Society of London, 10*, 275–310.

Akyürek, E. (1998). Fourth to eleventh centuries. In M. Sözen (Ed.), *Kapadokya* (pp. 227–395). Ayhan Şahenk Foundation.

Aldehuelo, N. (2003). Le Monastére byzantin d'erdemli. *Dossiers d'archeologie, 283*, 73–79.

Alıcı Şen, P., Temel, A., & Gourgaurd, A. (2004). Petrogenetic modelling of Quaternary post-collisional volcanism: A case study of central and eastern Anatolia. *Geological Magazine, 141*(1), 81–98.

Ambraseys, N., & Finkel, C. (1995). *The seismicity of Turkey*. Eren Press.

Andolfato, U., & Zucchi, F. (1971). The physical setting. In L. Giovannini (Ed.), *Arts of Cappadocia* (pp. 51–66). Nagel.

Arda, G. (2017). *A cave settlement in Ardahan (Turkey): Harosman.* Paper presented at the HYPOGEA, Cappadocia, Turkey.

Aydar, E., Schmitt, A., Çubukçu, H., Akın, L., Ersoy, O., Şen, E., Duncan, R. A., & Atıcı, G. (2012). Correlation of ignimbrites in the central Anatolian volcanic province using zircon and plagioclase ages and zircon compositions. *Journal of Volcanology and Geothermal Research, 213–214*, 83–97.

Aydar, E., Şen, E., Sarıkaya, M., & Kuzucuoğlu, C. (2019). In the footsteps of Strabon: Mount Erciyes Volcano—The roof of Central Anatolia and Sultansazliği Basin. In C. Kuzucuoğlu, A. Çiner, & N. Kazancı (Eds.), *Landscapes and landforms of Turkey* (pp. 565–576). Springer Nature Switzerland AG.

Balkan-Atlı, N., Slimak, L., Kuhn, S., & Açıkgöz, F. (2007). *Kömürcü-Kaletepe Obsidyen Atölyesi Paleolitik Dönem 2006 Yılı Kazısı.* Paper presented at the 29. Kazı Sonuçları Toplantısı, Kocaeli.

Belke, K., & Mersich, N. (1990). *Phrygien und Pisidien* (Vol. 7). Österreichische Akademie der Wissenschaften.

Berndt-Ersöz, S. (2006). *Phrygian Rock-Cut Shrines*. Brill.

Bertucci, G., Bixio, R., & Traverso, M. (1995). *Le città sotterranee della Cappadocia*. Erga Edizioni.

Bixio, A., Bixio, R., & De Pascale, A. (2023). *Göreme and Şahinefendi; History inside the rocks of Cappadocia*. Archaeopress Publishing Ltd.

Bixio, R. (Ed.). (2012). *Cappadocia: Records of the underground sites*. Archaeopress.

Bixio, R., Caloi, V., Castellani, V., & Traverso, M. (2009). *Ani 2004, surveys on the underground settlements*.

Bixio, R., Castellani, V., & Succhiarelli, C. (2002). *Cappadocia, le città sotterranee*. Istituto Poligrafico e Zecca dello Stato.

Bixio, R., & De Pascale, A. (2011). *Ahlat 2007, preliminary surveys on the underground structures*.

Bixio, R., De Pascale, A., & Karamağaralı, N. (2013). *Ahlat 2008, Second campaign of surveys on the underground structures*.

Bixio, R., De Pascale, A., & Karamağaralı, N. (2014). *Ahlat 2009, Third campaign of surveys on the underground structures*.

Bixio, R., De Pascale, A., & Karamağaralı, N. (2015). *Ahlat 2010, Fourth campaign of surveys on the underground structures*.

Bixio, R., De Pascale, A., & Pektaş, K. (2011). Indagini preliminari sulle strutture ipogee del Kale di Bitlis (Turchia sud-orientale). *Archeologia Medievale, XXXVIII*, 321–332.

Çevik, N. (2000). *Urartu Kaya Mezarları ve Ölü Gömme Gelenekleri*. Ankara. Türk Tarih Kurumu Yayınları No: VI-58.

Chantre, E. (1898). *Recherches Archeologiques dans l'Asie Occidentale: Mission en Cappadoce, 1893–1894*.

Darga, M. (1998). Second millennium B.C.: Middle and late bronze age. In M. Sözen (Ed.), *Kapadokya* (pp. 125–169). Istanbul: Ayhan Şahenk Foundation.

De Pascale, A., & Bixio, R. (2010). *Underground passage in defensive structures of Eastern turkey: The cases of Bitlis, Ahlat and Ani*. Paper presented at the Rupestrian Settlements in the Mediterranean Region, Massafra, Palagianello.

De Tchihatcheff, P. (1853–1860). *Asie Mineure: Description physique, statistique et archeologique de cette contree*. Gide et J. Baudry.

Durukan, M. (2012). *Kappadokia'da, Argaios Dağı Çevresinde Hellenistik-Roma Dönemi Mezarları ve Ölü Kültü (Graber und Totenkult In Der Hellenistisch-Römischen Zeit In Der Umgebung Des Argaios In Kapadokien)*. Istanbul. Arkeoloji ve Sanat Yayınları.

Equini Schneider, E. (1997). Varia Cappadocia. *Archeologia classica, 49*, 101–209.

Erkiletlioğlu, H. (2019). *Kappadokia Krallığı ve Roma-Bizans Dönemi*. Kayseri Kayseri Büyükşehir Belediyesi.

Esin, U. (1998). Paleolitik'ten İlk Tunç Çağı'nın Sonuna: Tarihöncesi Çağların Kapadokya'sı. In M. Sözen (Ed.), *Kapadokya* (pp. 62–123). Ayhan Şahenk Foundation.

Gilli, E., Yamaç, A., & Tok, E. (2014). Halys deviation tunnel and cliff dwellings of Sarıhıdır (Cappadocia–Turkey). *Opera Ipogea, 1*, 29–36.

Günay, G. (2024). Byzantine rock-cut architecture in Cappadocia and beyond: The state of scholarship. *Valonia, 1*, 35–63.

Haldon, J. (2024). Cappadocia: Old questions, new approaches—A historian's perspective. *Valonia: A Journal of Anatolian Pasts, 1*, 17–33.

Hamann, Y., Wulf, S., Ersoy, O., Ehrmann, W., Aydar, E., & Schmiedl, G. (2010). First evidence of a distal early Holocene ash layer in Eastern Mediterranean deep-sea sediments derived from the Anatolian volcanic province. *Quaternary Research, 73*(3), 497–506.

Hamilton, W. (1842). *Researches in Asia Minor, Pontus and Armenia: With some account of their antiquities and geology*.

Haspels, E. (1971). *The highlands of Phrygia: Sites and monuments*. Princeton University Press.

Herodotus. (1920). *The Histories* (Vol. 1). Harvard University Press.

Hild, F., & Restle, M. (1981). *Kappadokien (Kappadokia, Charsianon, Sebasteia und Lykandos)*. Verlag der Österreichischen Akademie der Wissenshaften.

Innocenti, F., Mazzuoli, R., Pasquarè, G., Radicati Di Brozolo, F., & Villari, L. (1975). The Neogene calcalkaline volcanism of Central Anatolia: geochronological data on Kayseri-Nigde area. *Geology Magazine*, 112.

Jerphanion, G. D. (1925–1942). *Une nouvelle province de l'art byzantin, les églises rupestres de Cappadoce*.

Jolivet-Lévy, C. (2015). *La Cappadoce : un siècle après G. de Jerphanion*. Geuthner.

Jones, H. (1929). *The geography of Strabo* (Vol. VI, Books 12-13-14). London.

Kalas, V. (2004). Early explorations of Cappadocia and the monastic myth. *Byzantine and Modern Greek Studies, 28*, 101–119.

Kalas, V. (2019). *The fortification wall at Selime*. Paper presented at the Space and Communities in Byzantine Anatolia, The Fifth International Sevgi Gönül Byzantine Studies Symposium, Istanbul.

Ketin, I. (1995). Erciyes Dağı volkan topluluğunun jeolojik evrimi. The geological evolution of the volcanic complex of Mt. Erciyes. In O. Guler (Ed.), *Antik çağ Ikonografisinde Erciyes*. Arkeoloji ve Sanat Yayınları

Kılıç, Ş., & Bayrı, B. (2021). *Byzantine Studies in Turkey: A bibliography (19th century–2020)*. Vehbi Koç Vakfı.

Kiepert, H. (1903). https://commons.wikimedia.org/wiki/File:Heinrich_Kiepert._Asia_citerior.Cappadocia.jpg. Retrieved 4 Aug. 2021.

Kipshidze, A. (1972). *Peshchery Ani*. Erevan: Academy of Sciences of the Armenian SSR Institude of Archeology and Ethnography.

Kostof, S. (1972). *Caves of god: The monastic environment of Byzantine Cappadocia*. MIT Press.

Kuhn, S., Balkan-Atlı, N., & Dinçer, B. (2009). 2008 Excavations at Kaletepe Deresi 3. *Anatolia Antiqua, XVII*, 291–299.

Kulakoğlu, F. (2011). Kültepe-Kaneš: A second millennium BCE trading center on the central plateau. In G. McMahon & S. Steadman (Eds.), *The Oxford handbook of Ancient Anatolia: (10,000–323 BCE)*. Oxford University Press.

Kulakoğlu, F., & Kangal, S. (2011). *Anatolia's Prologue: Kültepe Kanesh Karum*.

Kuzucuoğlu, C., Özbaşaran, M., Dumoulin, J., & Saulnier-Copard, S. (2020). Geoarchaeological and chronological reconstruction of the Aşıklı PPN site spatial development (Central Anatolia, Turkey). In Z. Tsirtsoni, C. Kuzucuoglu, P. Nondedeo, & O. Weller (Eds.), *Different times? Archaeological and environmental data from intra-site and off-site sequences* (pp. 43–63). Archaeopress Publishing Ltd.

Kürçer, A., Avcu, İ., Güler, T., Yıldırgan, M., & Gürsoy, H. (2024). *Kayseri Yöresinin Diri Faylarının Aktif Tektonik ve Paleosismik Özellikleri*. MTA.

Le Pennec, J., Bourdier, J., Froger, J., Temel, A., Camus, G., & Gourgaud, A. (1994). Neogene ignimbrites of the Nevsehir Plateau (Central Turkey), stratigraphy, distribution and source constraints. *Journal of Volcanology and Geothermal Research, 63*, 59–87.

Le Pennec, J., Temel, A., Froger, J., Sen, S., Gourgaud, A., & Bourdier, J. (2005). Stratigraphy and age of the Cappadocia ignimbrites, Turkey: Reconciling field constraints with paleontologic, radiochronologic, geochemical and paleomagnetic data. *Journal of Volcanology and Geothermal Research, Elsevier, 141*, 45–64.

Lucas, P. (1712). *Voyage du Sieur Paul Lucas, fait par ordre du roy dans la Grece, l'Asie Mineure, la Macedoine et l'Afrique* (Vol. 1). Nicolas Simart.

Mellaart, J. (1967). *Çatal Hüyük: A Neolithic Town in Anatolia*. Thames and Hudson.

Mitchell, S. (2018). Cappadocia. In O. Nicholson (Ed.), *The Oxford dictionary of late antiquity* (p. 279). Oxford University Press.

Mora, C., Balza, M., Bixio. R., De Pascale, A. (2020). *Cappadocia, a "Hidden" Landscape: "Underground World" and Neo-Hittite Evidence*. Paper presented at the 11th International Congress on the Archaeology of the Ancient Near East, Munich.

OBRUK. (2012). *Gaziantep Yeraltı Yapıları Envanteri 2012 Raporu*. Retrieved from Gaziantep.

OBRUK. (2019). *Fırat Nehri Kaya Yerleşimleri Envanteri*. Retrieved from Istanbul.

Ousterhout, R. (2011). *John Henry Haynes. A Photographer and Archaeologist in the Ottoman Empire 1881–1900*. Cornucopia.

Ousterhout, R. (2017). *Visualizing community: Art, material culture, and settlement in Byzantine Cappadocia* (Vol. 46). Harvard University Press.

Özbaşaran, M., & Duru, G. (2011). Aşıklı Höyük un Village d'Anatolie Centrale, il ya 10.000 Ans. *Archeologia, 489*, 54–65.

Öztürk, F. (2017). Rock-cut Architecture. In P. Niewöhner (Ed.), *The archaeology of Byzantine Anatolia: From the end of late antiquity until the coming of the Turks* (pp. 148–159). Oxford University Press.

Ramsay, W. (1890). *The historical geography of Asia minor*. John Murray.

Ramsay, W. (1897). *Impressions of Turkey during twelve years' wanderings*.

Ramsay, W., & Bell, G. (1909). *The thousand and one Churches*.

Restle, M. (1967). *Byzantine wall painting in Asia minor*. New York Graphic Society.

Rosada, G., Lachin, M., & Turchetto, J. (2024). Cappadocia. In B. Burrell (Ed.), *A companion to the archaeology of the Roman empire* (Vol. 2). John Wiley & Sons, Inc.

Rott, H. (1908). *Kleinasiatische Denkmäler aus Pisidien*. Kappadokien und Lykien.

Sagona, A., & Zimansky, P. (2009). *Ancient Turkey*. Routledge.

Schmitt, A., Danisik, M., Aydar, E., Lovera, O., Şen, E., & Ulusoy, I. (2014). Identifying the volcanic eruption depicted in a neolithic painting at Catalhoyuk, Central Anatolia, Turkey. *PLoS ONE, 9*(1).

Slimak, L., Kuhn, S., Roche, H., Mouralis, D., Buitenhuis, H., Balkan-Atlı, N., Binder, D., Kuzucuoglu, C., & Guillou, H. (2008). Kaletepe Deresi 3 (Turkey): Archaeological evidence for early human settlement in Central Anatolia. *Journal of Human Evolution, 54*(1), 99–111.

Sözen, M. (1998). *Kapadokya*. Ayhan Şahenk Vakfı.

Strzygowski, J. (1901). *Orient oder Rom: Beitrage Zur Geschichte der Spatantiken und Fruhchristlichen Kunst*. J.C.Hinrichs'sche Buchhandlung.

Strzygowski, J. (1903). *Kleinasien, ein Neuland der Kunstgeschichte*. J.C.Hinrichs'sche Buchhandlung.

Şen, E., Kürkçüoğlu, B., Aydar, E., Gourgaud, A., & Vincent, P. (2003). Volcanological evolution of Mount Erciyes stratovolcano and origin of the Velibaba Tepe ignimbrite (Central Anatolia, Turkey). *Journal of Volcanology and Geothermal Research, 125*, 225–246.

Tatar, O., Piper, J., & Gürsoy, H. (2000). Palaeomagnetic study of the Erciyes sector of the Ecemis fault zone; neotectonic deformation in the southeastern part of the Anatolian Block. *Geological Society Special Publications, 173*, 423–440.

Tekin, O. (1998). The kingdom of Cappadocia during the Hellenistic and Roman Times. In M. Sözen (Ed.), *Kapadokya* (pp. 195–227). Ayhan Şahenk Foundation.

Texier, C. (1839). *Description de l'Asie Mineure faite par ordre du Governement François de 1833 a 1837*. Didot Frères.

Thierry, N., & Thierry, M. (1963). *Nouvelles églises rupestres de Cappadoce*. Région du Hasan Dagi.

Toprak, V., & Göncüoğlu, M. (1993). Tectonic control on the development of the Neogene-Quaternary Central Anatolian Volcanic Province, Turkey. *Journal of Volcanology and Geothermal Research, 28*(3–4), 357–369.

Tozer, H. (1881). *Turkısh Armenia and Eastern Asia minor*. London Longmans, Green And Co.

Triolet, J., & Triolet, L. (1993). *Les villes souterraines de Cappadoce*. Torcy.

Tryon, C., Logan, M. A., Mouralis, D., Kuhn, S., Slimak, L., & Balkan-Atlı, N. (2009). Building a tephrostratigraphic framework for the Paleolithic of Central Anatolia, Turkey. *Journal of Archaeological Science, 36*(3), 637–652.

Uçkan, B. Y. O. (2010). *Rock architecture in Phrygia region in the Byzantine period*. Eskişehir.

Urban, M. (1973a). Das Rätsel der unterirdischen Städte Südostanatoliens. *Dritter Teil: Maginotlinie der Frühgeschichte. Vorland, 8*, 205–212.

Urban, M. (1973b). Das Rätsel der unterirdischen Städte Südostanatoliens. *Erster Teil: Der Befund. Vorland, 6*, 150–153.

Urban, M. (1973c). Das Rätsel der unterirdischen Städte Südostanatoliens. Zweiter Teil: Geschichtlicher Rahmen und Deutungen. *Vorland, 7*, 174–181.

White, G. (1904). The cavate dwellings of Cappadocia. In H. Baum (Ed.), *Records of the past* (Vol. 3, pp. 67–73). Records of the Past Exploration Society.

Yakubovich, I. (2014). *From Lower Land to Cappadocia*. In M. Kozuk (Ed.) Extraction and Control: Studies in Honour of Matthew W. Stolper. Oriental Institute.

Yamaç, A. (2012). Gödet'ten Taşkale'ye. *OBRUK, 5*, 22–24.

Yamaç, A. (2013). *Artificial cavities of Gaziantep*. Paper presented at the 16. International Congress of Speleology, Brno.

Yamaç, A., & Gilli, E. (2018). *Rock dwelled Churches of Sille*. Paper presented at the 4. MESS Symposium, Antalya.

Yamaç, A., & Okuducu, Z. (2017). *Underground hydraulic structures of Gaziantep (Southeast Turkey): Livas and Qastels*. Paper presented at the HYPOGEA, Cappadocia, Turkey.

Yıldırım, C., Sarıkaya, M., & Çiner, A. (2016). Late Pleistocene intraplate extension of the Central Anatolian Plateau, Turkey: Inferences from cosmogenic exposure dating of alluvial fan, landslide, and moraine surfaces along the Ecemiş Fault Zone. *Tectonics, 35*(6).

Yörükoğlu, Ö., Sevil, T., Taşcı, Z., & Türkmen, K. (1989). *Underground Cities in Cappadocia*.

Underground Shelters of Cappadocia

3

Abstract

This chapter is the core of the book, in which the underground shelters of Cappadocia are examined in comprehensive detail and evaluated from a variety of perspectives. Following an initial overview, the opening section examines the various typologies of these shelters and the standards associated with them. The date of the first construction of these structures is a topic of contention in numerous articles. The third section will present these disparate views and engage in a discussion of the dating of these structures. It is of equal importance to consider the defensive devices within the underground shelters and the defensive designs of these structures, which will be discussed in the following section. The subsequent section examines the techniques utilized for subterranean excavation in Cappadocia, with a particular focus on the methodologies employed in the excavation of these structures. The sixth and final section presents an inventory of 253 underground shelters that have been identified within the region to date.

Keywords

Underground shelter · Underground city · Artificial cavity · Cappadocia

3.1 Introduction

Although Ramsey, White, and Jerphanion alluded to the presence of millstone doors in the tunnels of underground shelters in their respective writings in 1897, 1904, and 1927 (Jerphanion, 1925–42; Ramsay, 1897; White, 1904), and these structures were also referenced by other authors until the 1960s, the first comprehensive study on Cappadocia's underground shelters was the article by Mahmut Akok and Hikmet Gürçay, published in the Türk Arkeoloji Dergisi in 1965 (Fig. 3.1) (Akok & Gürçay, 1965).

Akok and Gürçay have described Göstesin (Ovaören), Kaymaklı, and Derinkuyu shelters in their articles, and have drawn many sketches as well as the plan of Ören Underground Shelter. At the time this article was written, Kaymaklı and Derinkuyu were open to tourism, although a proper plan had not yet been drawn. Akok and Gürçay have provided very general information about these two shelters, but have described in detail the structure in Göstesin that has not been investigated before. The last paragraph of the article is as follows:

> Since it is understood that these underground facilities we have seen have narrow corridors and secure doors from all sides leading to other sections, we understand that the underground facilities of Göstesin have wider sections. We believe that in order to definitively reveal the underground facilities here, an intervention with a scientific method is necessary in the future. (Akok & Gürçay, 1965)

Martin Urban's three articles published in Vorland in 1973 were the first to provide a general comment on the underground shelters in Cappadocia, attempting to evaluate these structures as a whole rather than examining them individually (Urban, 1973a, 1973b, 1973c). This approach would not be seen again until the National Commission for Artificial Cavities of the Italian Speleological Society's project in the 1990s.

In his articles, the author focused on the usage, construction purposes, and most importantly, dating of the shelters rather than the technical and architectural details of these structures. Urban's hypotheses on the dating of shelters will be discussed in detail in Sect. 3.3.

Actually, the initial publication concerning Cappadocia underground shelters was Martin Urban's (1969) book, Yeralti Schechri (Yeralti Schechri/Yeraltı Şehri—Underground City in Turkish). But, this work was published privately, with only 120 copies printed, which limited its scope for wider evaluation. However, it included the Örentepe Shelter plan (Fig. 3.2) and Urban explored 12 other shelters in the Nevşehir area (Urban, 1969).

A. Yamaç and E. Tok, *Underground Shelters of Cappadocia, Turkey*, Advances in Karst Science,
https://doi.org/10.1007/978-3-032-16095-9_3

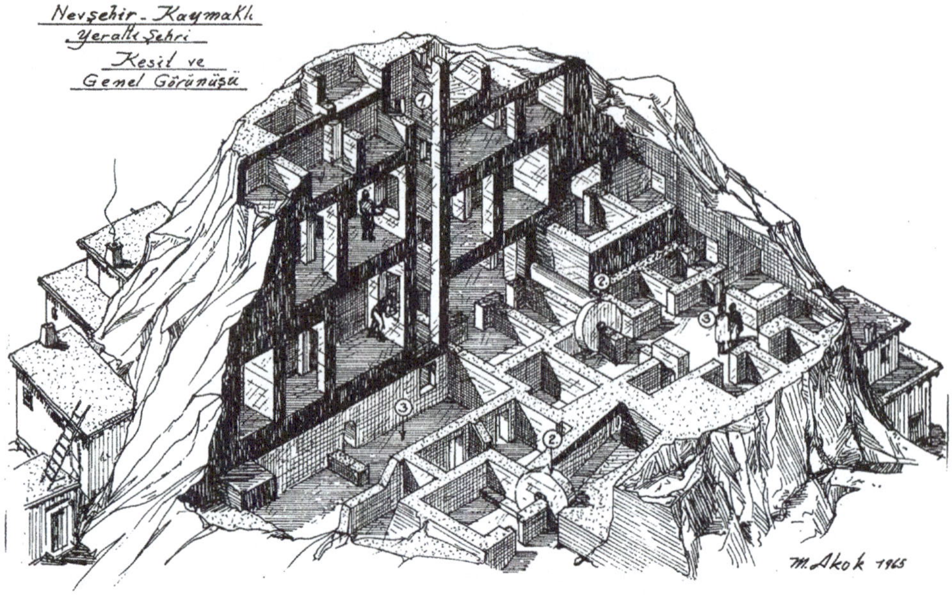

Fig. 3.1 3D sketch of Kaymaklı underground shelter by Mahmut Akok (Akok & Gürçay, 1965)

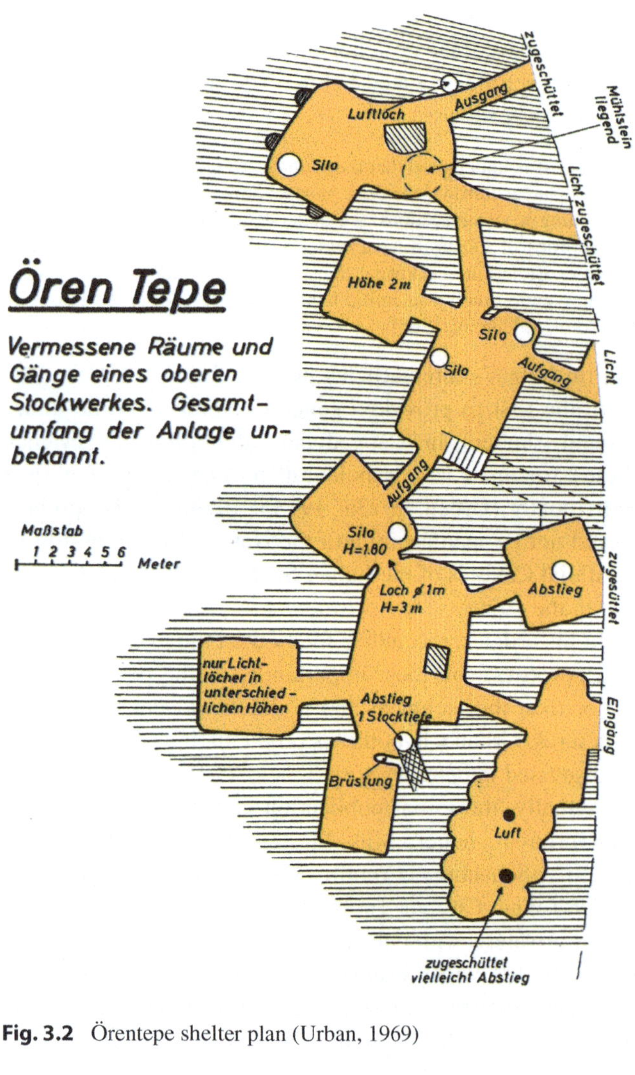

Fig. 3.2 Örentepe shelter plan (Urban, 1969)

Published in 1988 by Ö. Yörükoğlu, T. Sevil, Z. Taşcı, and K.T. Türkmen, "Les Sites Souterrains de Cappadoce" is a small tourist booklet, but it can be considered the first database on the underground shelters of Cappadocia in terms of the lists it contains. On the other hand, most of the information in the book is incomplete or incorrect, and most of the shelter plans are rough sketches (Yörükoğlu et al., 1988).

The project, conducted in the 1990s by the National Commission for Artificial Cavities of the Italian Speleological Society in collaboration with the Turkish Ministry of Culture and the Archaeological Museum of Nevşehir, constituted the inaugural comprehensive survey on Cappadocian shelters, as it not only involved the exploration and documentation of rock and underground shelters in six provinces, but also comprised a detailed scientific study of several structures. In addition to the production of numerous articles, three books have been published as part of the project. These publications have facilitated the emergence of a previously unexplored aspect of Cappadocia (Bertucci et al., 1995; Bixio, 2012; Bixio et al., 2002a).

During this project, a total of 35 underground shelters were surveyed. Although some of these structures, such as Derinkuyu, Kaymaklı, or Özkonak, had previously drawn plans, these missing or incorrect drawings were corrected, and large shelters such as Sivasa were explored and surveyed in detail.

In 2014, The "Kayseri Underground Structures Inventory Project," started by the OBRUK Cave Research Group in partnership with the Kayseri Metropolitan Municipality and the ÇEKÜL Foundation, has led to the discovery, survey, and mapping of 43 subterranean shelters within this province. To date, 12 articles, 7 preliminary reports, and a book have been published regarding this ongoing project, which aims to inventory not only the shelters but also all underground and rocky structures in Kayseri Province.

Prepared using the results of these two projects and many other different sources, the three most comprehensive inventories of these shelters are "Cappadocia: Records of the Underground Sites" published in 2012 under the editorship of Bixio (2012), "The Archaeological Settlements of Turkey" (www.tayproject.org) (Akyürek et al., 2015) and "Artificial Cavities of Turkey" (Bixio et al., 2021).

We used the aforementioned inventories to compile a list of the existing underground shelters in Sect. 3.6. This list includes 253 structures whose names and locations have been confirmed. In contrast, as we mentioned above, only 90 of these structures have been the subject of scientific research and survey.

Given the existence of at least one underground shelter in each of Cappadocia's villages, and in some cases multiple shelters, it seems reasonable to propose that the region should have a greater number of such structures than it currently does. The area in question covers 50,000 km^2, comprising six provinces, yet there are only a limited number of underground structures. This discrepancy can be attributed to a lack of comprehensive research, and it is likely that the number of such structures will increase in the future.

Underground shelters serve as enduring symbols of the troglodyte culture in the region, representing the tangible legacy of the Cappadocian people's resilience and ingenuity in a specific historical epoch. The labor observed at every stage of the underground shelter construction process, from initial planning to final excavation, serves as a proof of the challenging conditions that the region's inhabitants have endured for centuries.

On the other hand, due to lack of research and exploration in this area, it is not surprising that certain questions remain unanswered and subject to dispute. The disagreements on classification, terminology, date, and methodologies increase the complications, indicating the need for additional academic investigation into these enigmatic structures. Not only underground shelters, but all other rock-cut structures except churches are part of these discussions.

The underground and rock-cut structures of Cappadocia are remarkable in both quantity and diversity. Various types of rock-cut constructions exist, including those utilized as dwellings, warehouses, barns, apiaries, and dovecotes. Numerous rock-cut hydric structures also exist, including irrigation tunnels and cisterns, alongside religious edifices like churches, monasteries, and tombs, as well as defensive structures like underground shelters.

There are many studies on the classification or typology of rock-cut and underground structures, but such classification is very complex and difficult. For example, although the "typological tree of artificial caves" in the article published by the Italian Speleological Federation, Artificial Cavity Commission is extremely detailed, it is very difficult to apply to Cappadocia (Galeazzi, 2013). Yet, taking into account Cappadocia, the typology proposed by Bixio for the functional and architectural classification of all the structures identified so far in this region, which divides these structures into 12 different groups, is partially sufficient for the time being (Bixio, 2012):

These artificial cave groups are

- Cone dwellings/village
- Cliff dwellings/village
- Rocky castle-village
- Rocky courtyard settlement
- Rocky monastery
- Rocky church
- Rocky tomb
- Rocky dovecote
- Rocky apiary
- Underground city
- Underground shelter
- Hydric systems.

However, in a region such as Cappadocia where troglodyte life continues uninterrupted and is ongoing, even such a simple and easy classification of rock structures can become controversial. For instance: What is the reference for typology? Is it the initial purpose or the final state of the structure? There are many examples where a rock-cut structure carved as a church in the tenth century was transformed into a dovecote at some indefinite time.

3.2 Underground Shelters

Among the 253 underground shelters identified in Cappadocia thus far, those subjected to thorough examination exhibit a variety of architectural styles. These structures, irrespective of their size, display various typologies shaped by geographical differences and the geological properties of the rock from which they are carved.

These underground structures are generally located in open zones of the upland plain, where hiding places are not easily found. They are excavated directly beneath the earth's surface or, in certain instances, into elevated landforms such as buttes, cliffs, or hills.

These structures often extend deep into the rock, as they may consist of a single level or multiple levels, with all entrances located at the first and most superficial of those, as well as various chambers interconnected through a network of tunnels attributed as the underground road system and associated facilities. A distinctive defensive feature of these settlements is the presence of specialized security mechanisms, such as "millstone doors," which allowed for the isolation and fortification of extensive sections of the underground complex.

The defensive structures of Cappadocia can be classified from multiple typological perspectives; one key distinction is between an "underground shelter" and an "underground city." Although no definitive consensus has been reached, differentiating between "shelters" and "cities" is crucial in considering their distinct construction phases.

3.2.1 Underground City vs Underground Shelter

The terminology surrounding underground shelters remains a topic of debate. While the term "underground city" is widely used for tourism-related purposes, its precise definition is contested within the academic community. Various scholars have indiscriminately referred to all such subterranean defensive structures as "underground cities" (Ayhan, 2004; Gülyaz & Yenipınar, 2007; Ousterhout, 2017; Triolet & Triolet, 1993; Yörükoğlu, 1989). However, the appropriateness of labeling vast interconnected complexes like Derinkuyu, which originated as small defensive structures beneath dwellings and gradually expanded into large, interconnected systems beneath entire villages, as an "underground city"—the same term used for a small defensive structure with only a few rooms and short tunnels—raises questions. Regarding this debate, Roberto Bixio offers a refined classification based on the spatial distribution and relation of architectural spaces with different functions, as distinguishing between interconnected structures, which he terms "underground cities," and those that are more compact in nature, which he categorizes as "underground shelters" (Bixio, 2012).

Although the proposal put forward by Bixio is logical from a technical and architectural standpoint, and the two distinct terms proposed for these structures have their own consistent and logical aspects, it is inevitable that controversial situations will arise in practice.

In accordance with this definition, it can be stated that not only Derinkuyu but also several other defense structures should be accepted as "underground cities." In fact, we can go even further and say that although not all of them have been studied in detail, a large part of the defensive structures located under the villages in Cappadocia are composed of singular structures that have been connected with each other over time, and that they should be defined as "underground cities" because of these structural characteristics.

On the other hand, large but structurally independent defensive structures such as Sivasa and Filikören conform to the definition of a shelter due to their architectural characteristics as singular entities. The details of this typological classification are further elaborated below.

The defining feature that separates an underground city from a shelter is the presence of numerous "satellite systems" within the former. The combination of numerous, independently excavated defensive structures beneath residential areas has resulted in the formation of a vast network.

Figure 3.3 shows a synthesis of the probable excavation sequence for an underground city:

- Phase 1: Independent excavation of several shelters.
- Phase 2: Radial expansion of tunnels and rooms across multiple levels.
- Phase 3: Connection of separate systems and implementation of defensive mechanisms.

This organizational structure, which also fulfilled a sheltering function, allowed inhabitants to navigate the complex during enemy invasions and facilitated counterattacks through alternative routes.

In these subterranean networks, interconnections among individual structures are not necessarily aligned with deep vertical shafts, as seen in Bixio's model of Derinkuyu (Bixio, 2012, Fig. 2.17). Additionally, no deep shaft exists in any part of the Derinkuyu complex, with the exception of touristic sector.

If it's accepted that an underground city develops through the gradual connection of multiple "satellite systems" or sectors, then most structures fitting this description

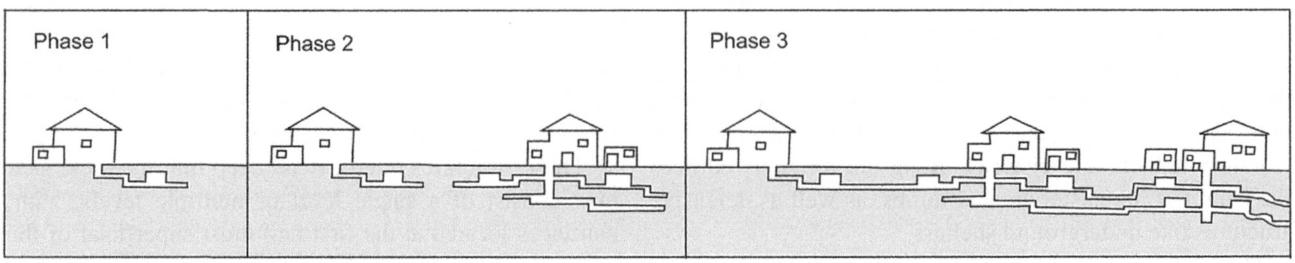

Fig. 3.3 The formation stages of an underground city (drawn A. Yamaç)

exhibit a cobweb-like configuration and are relatively shallow. Besides Derinkuyu, other examples meeting this definition include the defensive complexes at Büyük Bürüngüz, Kaymaklı, and Mimar Sinan, though they have not been fully explored.

We can assume that the development of these "singular" or "satellite" structures and their interconnection into a network took place over a long period of time. The gradual expansion of these structures likely stemmed from the realization that small defensive units were inadequate against prolonged raids. Over time, tunnels were extended and deepened, and additional millstone doors were incorporated.

Derinkuyu, the largest and best known of these underground cities, is described in detail in Chapter 4. In general, it is clear that this site consists of many sectors. The best known system is the one that is partly open to the public, which develops in a kind of helicoid structure around a central shaft that is intercepted many times at different levels. Three main overlapping belts can be recognized, with intermediate levels that go down to a depth of about 50 m. In the first one, just below the ground level, all the entrances to the underground net are located, generally hidden inside masonry buildings. From here you can go down to the second belt, through descending corridors, defended by millstone doors, where there are many rooms excavated in sequence, that may be considered as the "residential area." From this area other tunnels start, steep and narrow, with a lot of sudden changes in direction just before the operating rooms for millstone doors that defend the rooms in the deepest belt: the zone for the last shelter (redoubt) in case of raids. Other surrounding systems appear built in a similar way, and connected to each other through tunnels, now occluded or partially destroyed.

A comprehensive survey of Derinkuyu remains incomplete due to its sheer size, depth, and intricate layout (Aydan & Ulusay, 2013; Bixio et al., 2002b; Demir, 1990; Okuyucu, 2007).

3.2.2 Typology of Underground Shelters

The term "shelter" is used to describe hypogeal structures that were equipped with specific devices for the protection of the population from wars and raids. These structures have a long history in Cappadocia, with examples dating back centuries. The following sections provide a detailed description of the design features, devices, and excavation techniques of these widely dispersed yet often overlooked structures.

The architectural structures of shelters situated in close proximity to one another or in similar geographical formations frequently display similarities. For example, the underground shelters situated in the wide plain extending southwest of Gülşehir, Göstesin, Filiktepe, Sivasa, and Tatlarin have been found to exhibit strikingly similar characteristics (Bixio, 2012; Bixio et al., 2002a).

The excavations have been conducted within the confines of low, flat hills, colloquially referred to as "buttes." At the foot of the modest cliffs, the entrances to large rooms are visible, exhibiting indications of use as storage facilities and shelters for domestic animals, as well as churches. These compartments are connected to narrow tunnels that open into the tuff, leading toward the interior of the butte.

On the other hand, as observed in many different villages in the region, there are also small, detached shelters dug under the houses or in the wall adjacent to the rock. The entrances are invariably protected by one or more millstone doors. The investigations of the various underground systems have revealed the existence of a few independent subsystems, comprising tunnels dug with the opposite-front technique, which converge at a junction point. Each module was protected by multiple millstone doors, the efficacy of which was enhanced by additional safeguards, including right angles in the tunnels and bottlenecks.

It is noteworthy that wells extending to the water bed have been identified. The dimensions of each subterranean network are smaller than those of the so-called underground cities, yet they may exceed one kilometer in length. It seems plausible to suggest that this architectural style was a provisional refuge for a modest population and their livestock during incursions. One possible interpretation of such a complex structure is that the various family groups inhabited the more external rooms, stored food in the interior, and sought refuge in the redoubt when danger arose. In the event of necessity, they could provide assistance to each other through the long connecting tunnels and the emergency ways. Schematically we can divide shelters in the typologies hereafter described.

3.2.2.1 Horizontal Underground Shelters

The structures in question are essentially one-level constructions, exhibiting slight variations in altitude and the number of stories. They may be simple, with a linear configuration, or they may develop in interconnected networks. In general, these structures have few entrances, which are concealed among rocks or inside the first belt's cavities. These entrances open at the bottom of rocky crags. In the innermost part of the settlement, one or more rooms are located. These rooms serve a variety of purposes, including housing, and in times of conflict, they can be used as a refuge. The rooms are protected by millstone doors, which provide a form of defense against intruders. Additionally, the settlement features an internal water well and a dovecote, which can be utilized as the last line of defense for the inhabitants, or "redoubt," protected by a shaft-trap. The settlement also includes progressive defenses, such as wells and silos, which provide a source of sustenance and protection for the local population.

Sometimes the horizontal shelters are combined with rock-cut settlements, as cliff or terraced villages. In this case, the entry point to the part with defense devices, that of the part penetrating deep through the rocky mass up to the redoubt, is concealed inside a system of rooms for residential use excavated in the first belt. The entry points can be located at the ground level or at a raised level, as for the rock-cut villages of Karşı Kilise, Tatlarin, or Güzelöz (Bixio, 2012; Yamaç & Tok, 2015).

A very similar structure, but with a special location, can also be found outside the volcanic tuff area, in the Çıralı Sinkhole site, near Karapınar, in Konya Province, west of the territory today considered as Cappadocia. Here a rock-cut village excavated in the vertical cliff of a large karstic sinkhole and a horizontal tunnel, 165 m long, protected with vertical traps and millstone doors, connects the settlement with a small valley in the middle of the plateau. It seems therefore that the tunnel could be an emergency route to escape, in case of attack, or to take from behind the potential assailants, coming out in a higher position than the village (Grek & Podruckaya, 2012).

In Cappadocia, there are combined-type shelters that exhibit both structural and defensive characteristics, which are typical of both underground shelters and rock settlements. The study of such complexes is of great importance for an understanding of the development of these shelters. In the absence of archaeological excavations, it is possible to determine the period of establishment of such complexes only when the structure of the shelter complex includes churches, tombs, or other objects that can be dated according to their external characteristics.

A number of experts have posited that the surface relief exerted a pivotal influence on the configuration of the Cappadocian cave shelter complexes. In particular, in locations with pronounced rock outcrops, the defensive structures of cave shelters were typically constructed along the vertical stone walls of gorges or the inclined surfaces of tuff cone-shaped remnants. Conversely, in areas where rock outcrops are scarce, shelters predominantly take the form of underground cities, structures constructed within the depths of the rock massif (Bixio, 2012). However, there are also combined-type shelters in Cappadocia that exhibit a structure and defensive devices characteristic of both the aforementioned types of cave complexes (Bobrovsky & Grek, 2015; Grek et al., 2010). The study of such complexes is crucial for comprehending the genesis of the cave shelters of Cappadocia. It is evident that the evolution of diverse cave systems adheres to analogous principles, progressing in stages from rudimentary forms to more sophisticated ones. Despite the occasional employment of disparate techniques, the resemblance in shelter morphology is likely a consequence of analogous historical circumstances, with the possibility that they originate from a shared era.

3.2.2.2 Vertical Underground Shelters

Vertical shelters can be ascending or descending. In some cases, ascending shelters extend within the rocky mass on two or more overlapped levels. They are interconnected by tunnels, ramps, or often vertical shafts with footholds which, normally, have themselves the function of traps. Usually the rooms designated as the last shelters (redoubts) are located on the higher level. Examples of such *shelters* include those found in Meskendir Valley (Bobrovskyy & Grek, 2015), Tekkaya, between Göreme and Çavuşin (Bobrovskyy & Grek, 2016), or Mazıkoy (Yamaç & Tok, 2015).

Different from these stepped structures, there are also ascending shelters that are excavated entirely vertically. These complexes are excavated with a predominantly vertical development of rooms (overlapping) and are grouped around an axis that is more or less central. This allows the excavated spaces within the rock mass of a single pinnacle to be maintained while the sequence of defense devices remains consistent with the broader strategy of the defensive system. A shelter of this type was found in the site of Golgoli, near Mustafapaşa Village, 17 km south of Ürgüp (Yamaç & Gilli, 2016). Another one is located in the upper Zemi Valley (Göreme), northeast of Karabulut Kilisesi (Bobrovskyy et al., 2015).

Vertical "descending" shelters exist which, on the contrary, develop from top to bottom, so from the ground level to the deepest rock areas, where redoubts are located. Mucur, in Kırşehir Province or Kırlangıç, in Kayseri Province are typical descending vertical shelters (Yamaç, 2022). Kırlangıç Valley Underground Shelter with its five-story structure starts from the top of the valley and goes down to the stream on the valley floor.

When vertical shelters consist of overlapped levels but with their extension limited to the belt parallel with the rock wall nearest the outside where openings overlook, we will consider them as "rock-cut," rather than underground. In some cases, shelters extend within the rocky mass on two or more overlapped levels (ascending shelters). They are interconnected by tunnels, ramps, or, often, vertical shafts with footholds which, normally, have themselves the function of defense (traps), in addition to the devices prepared at the first level. The redoubt, in the higher level, may also have had a dovecote function, or have been later converted. There are many examples in Göreme, but also in Ihlara Valley, Province of Aksaray (Bixio et al., 2023; Grek et al., 2010).

3.3 Dating the Underground Shelters

The determination of the origins of Cappadocia's underground shelters remains a contentious issue due to the limited availability of historical records and archaeological evidence. The dating of these structures is largely speculative, adding further complexity to their historical narrative. Rock-cut structures in this region have undergone continuous modification and reuse over the centuries, with some still serving functional purposes today. It is plausible that defensive structures were no exception to this pattern, having been initially excavated well before the Byzantine period and subsequently expanded or altered over time. Massive underground cities such as Filiktepe, Göstesin, Derinkuyu, Kaymaklı, and Bürüngüz could not have been excavated in a short timeframe (Bixio, 2012).

Even if these underground shelters originated prior to the Byzantine era, their uninterrupted use over centuries has made it exceedingly difficult to obtain clear archaeological findings. Mehmet Arif Bilici, who conducted excavation and cleaning work on 16 small underground cities in Oymalı Village, noted in his report that:

> It is not possible to encounter dense findings in the underground shelters which were used in every period (Bilici, 2002).

An illustrative example comes from the Kayseri-Dimitre rock-cut settlement, where an elderly villager recounted that a surveyed structure had been his late aunt's house and that he had spent his childhood there. Despite its recent residential use, two millstone doors were found in the tunnels of this dwelling, indicating its original defensive function before being repurposed as a residence and storage space.

Given their defensive purpose, it is reasonable to assume that these underground shelters were kept secret during their construction, which explains their absence from historical records. However, some historical sources do reference people living in rock-cut dwellings in Anatolia.

The first writer to mention subterranean settlements and a troglodyte civilization is Xenophon. In his Anabasis (Book IV, 5, 25), written around 400 BC, which describes the 10,000 Greek warriors' great march from Sardis to Babylon and back to Greece from the coast of Black Sea, on their way back to their homeland, writes as follows:

> The houses of these villages are underground: their entrance doors look like well mouths, but the internal rooms are large. The animals enter through passages purposely dug; men enter by means of staircases. There they breed goats, sheep, cows, chickens with their offspring; all the cattle are fed exclusively with hay (Brownson, 1922).

Although it has been stated by some researchers (Giovannini, 1971) that these underground dwellings mentioned by Xenophon may refer to Cappadocia, it is almost certain that this return path of Greek warriors did not pass through Cappadocia. Since Xenophon wrote that this place is close to the source of the Euphrates River, it is much more likely that the area he described was Ani (Kars Province) or Georgia (Bixio, 2012; Urban, 1973a). Anyway, this may prove that the underground dwellings were already a fact in Asia Minor at the end of the fifth century BC.

Regarding the underground structures in Cappadocia specifically, the oldest known historical reference dates to the second half of the tenth century AD. The historian Leo the Deacon, describing an invasion of Cappadocia by Nikephoros Phocas against the Arabs, wrote:

> .... Driving its chariot in the sign of the Bull, Nikephoros left Byzantium and crossed over to the opposite shore of Asia. Upon arrival in Cappadocia, whose people were formerly called troglodytes, because they dwelt in caves, hollows, and labyrinths, as if in dens and holes, he established his quarters there (Talbot & Sullivan, 2005).

Almost in the same era, Arab geographer Ibn Hauqal makes another reference to the same Cappadocian cave dwellings and their inhabitants, whom he also describes as "troglodytes" in his work "Kitab surat al-ard" (Hauqal, 1964).

Due to the continuous occupation of rock-hewn structures for millennia, establishing precise dates for their origins remains an immense challenge. Researchers argue that it is nearly impossible to scientifically determine the exact period of the initial excavation of rock settlements. Over time, a Roman rock-cut tomb may have been repurposed as a residence, then a storage facility or barn. The walls are destroyed or built, and these transformations obscure the original structure, leaving little trace of past modifications.

Occasionally, remnants such as rock decorations hint at previous uses, such as a warehouse that was once a Roman tomb or a church that later served as a dovecote. The same process applies to underground shelters, which may have evolved into dwellings. This widespread practice throughout Cappadocia complicates efforts to ascertain the original construction dates of these rock-cut structures.

As discussed in Chapter 1, no ancient documents have been discovered detailing the underground settlements of Cappadocia. Such records could have provided comprehensive descriptions of the techniques, organization, and purposes of these communities. In the absence of any extant documents, contemporary records pertaining to the extensive, articulated, and diversified array of subterranean structures excavated within the region's volcanic terrain have been crucial in providing a dynamic, complex, and nuanced depiction of the communities that inhabited Cappadocia.

Scholars unanimously acknowledge the lack of direct historical documentation:

> Because of the absence of a written record, it may never be possible to write a traditional history of the region, but it is possible to visualize the kinds of communities that once formed the living landscape of Cappadocia (Ousterhout, 2017).

> In dramatic contrast to its dearth of textual evidence, remains of Cappadocia's material culture are unrivalled in the Byzantine world. All the same, [...] most areas of settlement remain undocumented and poorly understood. (Ousterhout, 2017).

This region, as of the seventh century AD, became the easternmost point of the Byzantine Empire, boundary with the Sasanian Empire, and during each weak era of the Byzantine Empire, Persian armies attacked this region. During the occupation of Caesarea by the Persians in 611, the residents fled, most eluding death or capture, and the city was destroyed. Once a major world power, the Sasanian Empire had exhausted its human and material resources after decades of warfare against the Byzantine Empire. Between the years 628 and 632, the civil war started in the Sasanian Empire which quite weakened this empire, and the Arabic armies, which took advantage of this situation, started to attack Anatolia.

Arab raids, which continued until the eleventh century, had an identity of a kind of "looting raid" that was repeated nearly every summer, rather than an invasion of the region. First the Sasanian and after the Arabic raids against Cappadocia continued for approximately 400 years by speeding especially during the weak eras of the Byzantine Empire (Haldon & Kennedy, 2006) (Fig. 3.4).

After the Arab attacks, the region was faced with the Seljuk Turks' raids this time. Seljuk armies destroyed Melitene in 1057 and Sebasteia in 1059. In 1067 they captured Caesarea, burnt it to the ground, and massacred all the people. As a result, the city remained uninhabited for the next half-century. Four years later, in 1071, Byzantine Emperor Romanos IV came to the region with a big army to take control of Cappadocia. After his defeat in the Battle of Mantzikert, the whole region was completely occupied by the Seljuk Turks (Ash, 2006; Hild & Restle, 1981).

It is known that as a result of all these wars and raids, which almost continued for four centuries, first with the Persians, then with the Arabs and Seljuks, there was a breakdown of the civil and military order, and a colossal displacement of population in Cappadocia. Those centuries were probably very unkind to the people of Cappadocia, as the chaos of invasions further slackened the government control in an already chaotic landscape troubled by banditry and lawlessness (Cooper & Decker, 2012).

However, a number of hypotheses regarding the age of these structures have emerged as a result of the lack of historical sources and archaeological excavations. In an article, Urban stated that these structures might have been established during the Phrygian border clashes with the Tabal Kingdom in the ninth and eighth centuries BC. The same author, as a continuation of this hypothesis, claims that Cappadocia served as a buffer zone during the conflicts between the Phrygians and the Assyrians in the late eighth and early seventh centuries BC, and that the underground shelters there may have been excavated up during this time (Urban, 1973b, 1973c).

The area corresponding to Cappadocia, probably coinciding with the "Low or Lower Land" frequently mentioned in the Hittite texts, in the second millennium BC was located in the middle of their empire. Reasonably, it is identifiable as Tabal, one of the Neo-Hittite kingdoms in which the Empire shattered in the ninth to eighth centuries BC.

Some monumental Neo-Hittite hieroglyphic inscriptions of the kings of Tabal have been carved on rocky

Fig. 3.4 Arab dominance in the region during tenth century AD (base map HGM—elaboration A. Yamaç)

outcrops that are in the immediate vicinity of the entrance of artificial cavities, especially the inscriptions of Topada, Karaburna, Sivasa, and Yassı Höyük, all located in Western Cappadocia.

The Topada Monument, written in Luwian hieroglyphs from the Late Hittite period, is located near the village of Ağıllı in the Topada district of Nevşehir. The monument describes the military successes of the Tabal King Wasusarma and especially the fight against the city of Parzuta. It is dated to the second half of the eighth century BC (Hawkins, 2024) (Fig. 3.5).

The Karaburna Monument is located near the Karaburna village of Hacıbektaş, Nevşehir. It is on an agreement between "Sipi, the king" and "Sipi, the son of Ni" about the restoration of the fortress at the location. Inscription is likely to be also from the eighth century BC (Hawkins, 2024) (Fig. 3.6).

The idea that these monuments might be connected to nearby underground and rock-cut structures was first put forward by Urban, who even wrote that these underground defense structures could have been excavated in the first millennium BC. He also points out that perhaps the underground settlements not only had the defensive function, but also the function of holding together the fighting men. A little further on, the author addresses the issue of the first builders of these structures and hypothesizes that they were

Fig. 3.5 Topada monument (photo E. Soner)

Fig. 3.6 Karaburna monument (photo E. Soner)

the Phrygians, between the ninth and the eighth century BC, when they came into contact with the kingdom of Tabal (Urban, 1973b).

According to Urban, in the region of Tabal, the Phrygians would have found a suitable environment, and perhaps the presence of experts, possibly even qualified groups of tomb diggers able to realize large defensive fortifications. His map shows the position of most of the known underground sites of the Tabal area, which temporarily served as a "buffer zone" between the Phrygian and Assyrian kingdoms in Anatolia (Fig. 3.7). In support of this Phrygian hypothesis, Urban also mentions the inscriptions found in Tyana. He especially emphasizes that one of the two reported the name Mida and that the second one, found in 1900 and now lost, seems to have been engraved on a stone in the shape of a millstone.

In the light of the above considerations, Urban suggests that these structures should be backdated to the period of the conflicts between the Phrygians and the Assyrians (late eighth–earlyseventh centuries BC).

Based on these recent studies, it seems that the hypotheses of Urban do not lead to firm conclusions about the dating of the underground structures of Cappadocia. His suggestions, however, seem to be useful for further consideration on the subject.

We are particularly thinking about the observation that the underground refuges are for the most part placed in the area lying between the advance of the Phrygians from the west and of the Assyrians from the east between the eighth and the seventh centuries BC. During this period, this Luwian buffer zone was under the rule of the kings of Tabal, epigraphic and artistic evidence seems to demonstrate the existence of political, diplomatic, and military relationships between the inhabitants of Cappadocia and the Phrygian kingdom during the second half of the eighth century BC. Indeed, typical Phrygian motifs are on some of the most important monuments of the kingdoms of Tabal; other similarities between iconographic elements on the orthostates from Gordion and Ankara and contemporary Neo-Hittite art products in Cappadocia are noteworthy.

Although there are several hieroglyphic inscriptions at Topada, Karaburna, Sivasa, and the ruins of a walled city currently under excavation (Yassı Höyük), near underground shelters, so far it has not been possible to establish any direct link between them.

> this stage of research and knowledge, in absence of any archaeological proof, we can only speculate on this subject (Mora et al., 2017).

The same can be said about the hypothesis concerning the origin of the underground shelters between the eighth and seventh centuries BC. Martin Urban suggests that they could be considered actual military installations carried out at the time of the conflict between Phrygians and Assyrians, not just as shelters, but also for quartering troops ready for the deployment.

Urban claims that these structures, which were created much earlier, were later reused by Christian peoples. That opinion is because, indeed, if we display on a map the position of the underground sites so far identified, they seem to correspond to a strip of interposition that Urban calls "Maginotlinie der Fruhgeschichte," i.e., the "Maginot Line of the Ancient Times" (Urban, 1973a). However, with the current state of investigations, there is not tangible evidence that such structures, or parts of them, already exist at that distant historical period (Mora et al., 2020).

Despite all these hypotheses, the only concrete findings to date indicate that the underground shelters in Cappadocia were excavated by the local Christian population for self-defense and survival amid the tumultuous conditions widely between the seventh and eleventh centuries, a view supported by numerous researchers (Kempe, 1988; Triolet & Triolet, 1993, 2002).

Fig. 3.7 Map of Phrygian and Assyrian kingdoms around 670 BC (Wikimedia—elaboration A. Yamaç)

Fig. 3.8 Cross engraving on the wall of Golgoli underground shelter no 1 (photo A. Yamaç)

The walls of these shelters are engraved with dozens of crosses and the presence of a church in several of the underground Cappadocian shelters indicates that these structures can be dated to the Byzantine period (Bobrovskyy & Grek, 2018) (Fig. 3.8). The churches which were constructed right at the entrance of Ağırnas and St. Mercurius underground shelters and the tunnels continuing after these churches are protected with millstone doors show that these churches have also been used in normal times and during the raids, people have hidden inside. Also, there are small chapels in the depths of the Doğanlı and Danalık shelters of Kayseri Province (Yamaç, 2017; Yamaç & Tok, 2016; Yamaç et al., 2014) (Fig. 3.9).

Additionally, the archaeological excavations at Aydıntepe, Kavlaktepe, and Oymalı underground shelters revealed the presence of exclusively Byzantine pottery fragments, providing further evidence to support the aforementioned opinion (Bilici, 2002; Erkmen, 1998; Faydalı, 1992, 1993; Özkorucuklu, 1991; Şahin, 1993).

3.4 Mechanisms and Architectural Features of Underground Shelters

The underground shelters of Cappadocia are most renowned for their primary defensive feature—the millstone doors. However, these structures encompass a variety of additional mechanisms and architectural designs.

Recent comprehensive studies have yielded valuable insights into the fundamental aspects of these shelters, including their dimensions, spatial arrangements, excavation methodologies, and the strategic adaptations employed to improve both their practicality and security, extending beyond mere concealment.

Fig. 3.9 Church inside the Ağırnas underground shelter (photo R. Straub)

At present, these subterranean shelters, concealed within geological formations, are in a state of significant deterioration and often suffer from neglect. Nevertheless, they continue to be remarkable in their scale and meticulous organization. These hidden spaces, hewn into cliffs or below the earth's surface, were intended for regular use but were designed to be sealed off when necessary during times of crisis. They appear to have been specifically constructed to serve as temporary sanctuaries for either small groups or entire communities, offering protection against brief yet perilous threats.

The dimensions of these shelters vary greatly, ranging from relatively small chambers measuring only a few tens of meters to vast networks spanning over a kilometer in length. These intricate systems consist of numerous interconnected rooms connected by narrow passageways. Notable examples include the archaeological sites of

Filiktepe, Sivasa, Derinkuyu, and Büyük Bürüngüz, which, despite partial exploration due to collapses and sediment accumulation, extend beyond 1,500 m and contain multiple millstone doors.

Topographical assessments of these networks suggest that they were deliberately designed to optimize defensive effectiveness while also allowing for interaction among different groups of inhabitants.

Despite their unassuming external appearance, these underground shelters demonstrate advanced defensive mechanisms and architectural sophistication beyond the commonly recognized millstone doors. Between the seventh and eleventh centuries AD, the local population utilized these shelters to endure frequent Arab invasions, which occurred multiple times per year over nearly four centuries. It is plausible that these structures were designed for both defensive and offensive strategies.

This section examines the various mechanisms and architectural features of underground shelters based on the classification framework established by Triolet and Bixio, with certain modifications (Bixio & Castellani, 2002; Bixio & De Pascale, 2012; Triolet & Triolet, 2002). The primary alteration made was the decision not to classify these mechanisms strictly as either offensive or defensive, given that many fulfilled both roles simultaneously.

3.4.1 Millstone Doors

The predominant defense mechanism adopted in Cappadocia's underground and rock-cut settlements comprises securing the designated passageways with robust stone doors, referred to as millstone doors due to their circular form. These defensive elements were positioned at entrances to rooms and tunnels near the exterior, as well as at strategically significant interior locations, sometimes appearing in pairs within close proximity, regardless of the overall size of the shelter. This system not only functions to impede intruders from penetrating the shelter, but also may be utilized to entice them into predetermined traps. In a single explored shelter measuring 1,670 m in length, more than fifty such doors were identified, while even the smallest networks, consisting of just a few rooms, contained at least one.

Ramsay calls them "disc of stone":

> A big solid disc of stone stands in a niche, outside each underground house, ready to be drawn forward in front of the door at any alarm (Ramsay, 1897).

As far as we know, White is the first to compare them to millstone:

> When Ibrahim Pasha invaded Turkey half a century ago, with the Egyptian army, the villagers of Misli fled below ground, cutting off their rear by stone doors like millstones, which they rolled across the passageways (White, 1904).

Twenty years after White, Jerphanion would use the same word for these stone doors:

> Behind their millstones, the monastery dwellers could wait for Tabri's fire and dashes until the danger had passed (Jerphanion, 1925–42).

These devices are widespread over the whole territory of Cappadocia and beyond. This circumstance confirms that underground shelters and stone doors were part of a cultural and technical tradition common to all the population here resident (Triolet & Triolet, 2002).

In Anatolia, within rock-cut sites far from Cappadocia, as far as we currently know, only four underground shelters with millstone doors have been located: in the Köyiçi Shelter, 50 km northwest of Lake Tuz, Ankara Province, in Kalfat, Çankırı Province, northwest of Hattuşa, and in eastern Turkey, in Aydıntepe (Bayburt) and Hilar-Çayönü Tepe, 500 m southwest of Sesverenpınar (Ergani, Diyarbakir).

Although the tradition of using millstone doors in underground shelters is specific to Cappadocia, these devices have been widely used in many different regions for centuries to block the entrance to rock-cut tombs.

For example, 17 different rock-cut tombs have been found 500 m west of the Jupiter Dolichenus sacred site on Dülük Baba Hill, Gaziantep (Turkey), and all the tombs have in situ millstone doors. Here they were used as movable closures that allowed subsequent burials inside rock-cut tombs. Because of their proximity to the sacred location and fine workmanship, it is assumed that these graves belonged to Jupiter Dolichenus Temple priests. It is estimated that all of these tomb chambers, with extremely distinct designs, were excavated from the third century BC to the fourth century AD, spanning about 700 years (Ergeç, 2003; Wagner, 1982).

Similar doors have also been reported in Israel. Only five tombs, associated with wealthy families, are documented, dating from the first century BC to AD 70. Millstones with diameters ranging from 120 to 140 cm are positioned in designated ditches in front of the entrances. Millstones smaller in diameter, about 90 cm, and simply leaned over the tombs' opening, appear quite common between the late Roman and Byzantine period, between the second and seventh centuries AD (Kloner & Zissu, 2007).

Today, hundreds of millstone doors remain within Cappadocia's underground shelters in various states—fully open, semi-open, collapsed, or partially destroyed. In certain instances, they remain in the closed position or were abandoned during transport (Figs. 3.10 and 3.11). In many sites, millstones are lost and defensive devices are recognizable only thanks to the presence of operation rooms.

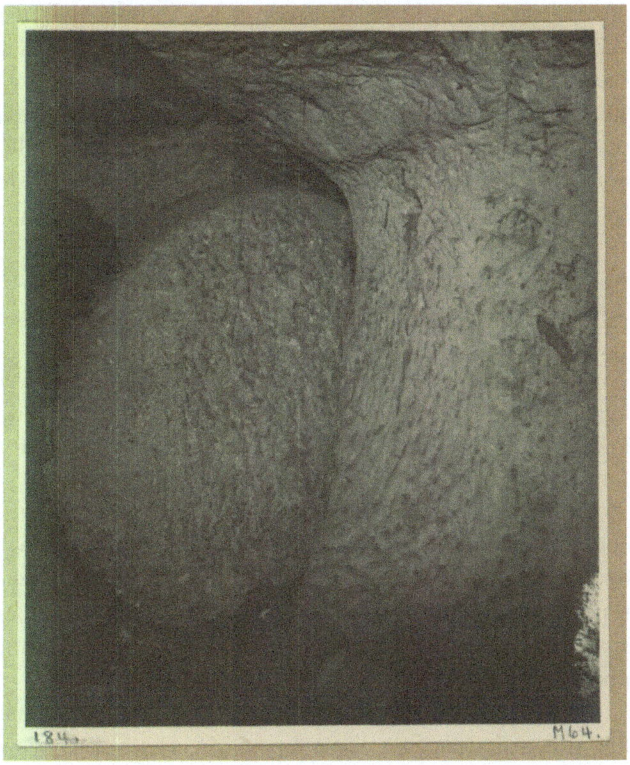

Fig. 3.10 One of the earliest photos of a millstone door in an unknown shelter, probably Kaymaklı, 1935 (photo J. D. Whiting)

Fig. 3.11 A millstone door abandoned during the transportation in Ötedere shelter no 2 (photo A.E. Keskin)

Millstone doors, normally, were excavated within the rooms they defended. A clear example is within the Orta Church at Ürgüp—Kayakapı. On one wall of this rock-cut church there is a roughly hewn and partially broken, unfinished monolithic millstone door (Fig. 3.12).

In some cases, these doors were fashioned from a harder rock type than the surrounding chamber walls, suggesting that builders sought out suitable rock outcrops for their construction before transporting them into underground complexes.

The diameter of millstone doors varies between 90 and 190 cm, according to the tunnel that had to be blocked, and the thickness is between 20 and 45 cm. In Büyük Bürüngüz Shelter, one of the total 27 millstone doors has a diameter of 185 cm and a thickness of 45 cm (Yamaç, 2019). There are three gigantic millstone doors in Ürgüp–Cappadocia, very close to each other. The millstone door in the Dereler Cliff Settlement has a diameter of 210 cm and a thickness of 67 cm, its weight must be around 5,500 kgs (Fig. 3.13).

However, they should not be difficult to handle from the inside even if two or four people were needed to push them, perhaps even with the help of a lever. Clearly, they were closed only in the imminence of an enemy attack. Once closed it would be extremely difficult to open them from the outside, if not impossible, and anyway a wedge was enough to make them unmovable by the assailants.

Fig. 3.12 An unfinished millstone door extracted from the main rock in Kayakapı Orta Church / Ürgüp (photo B. Yazlık)

The diameter of a millstone door naturally depends on the height of the tunnel it will cover, and according to a theory, the tunnels of the chambers that were utilized for

Fig. 3.13 A colossal millstone door in the Dereler cliff settlement—Ürgüp (photo A. Yamaç)

daily purposes, such as warehouses or storage rooms during peaceful times, were excavated higher for comfortable access. Consequently, the diameters of the millstone doors that covered these tunnels had to be proportionally large.

A thorough analysis by Bixio and Castellani investigated the millstone doors in four different shelters, culminating in the formulation of a graph that delineates diameter and thickness proportion (Bixio & Castellani, 2002). The graph reveals that the doors were manufactured within a specific ratio, with some exceptional deviations.

Most of these doors are provided with a central hole, which could serve as a spyhole and, probably, as a loophole to hit the intruders, becoming an active part in defense. It is possible that these holes were also used to simplify the handling of millstones, mostly when re-opened, for example by inserting a horizontal pole which could have facilitated the removal of the millstone at least from the groove, allowing a person to insinuate in that space and then to push. When the door has no spyhole, there are holes excavated in the ceiling or in the side walls of the tunnel that substitute the same function.

The handling of the millstone doors required the space necessary to roll the doors into a safe position, that isto avoid an easy overturning. To this purpose, the stone doors, once closed, have to be blocked. This can be achieved in various ways, the most common ones being by means of sockets in the walls and leaf pillars and/or slabs. The millstone doors are generally located either along a tunnel, in an operating room, or where a tunnel enters a larger room. It is not unusual in the case that the millstone door is preceded by a bent in the tunnel that could prevent the use of a ram.

3.4.2 Operation Rooms

To accommodate millstone doors, designated spaces were meticulously carved within the underground networks, incorporating support structures to prevent attackers from forcibly removing them.

These spaces, termed "operation rooms," were either compact chambers excavated along tunnels (enclosed operation rooms) or larger compartments within the shelter (open operation rooms). In both configurations, sufficient space was allocated to facilitate handling and structural support (Fig. 3.14).

In this room, a supporting pillar is utilized to maintain the millstone door in a vertical position. The pillar is hewn from the primary rock formation; alternatively, a substantial slab may serve a similar function. The objective is to create a passageway for the millstone door between the pillar and the wall of the room. This configuration created

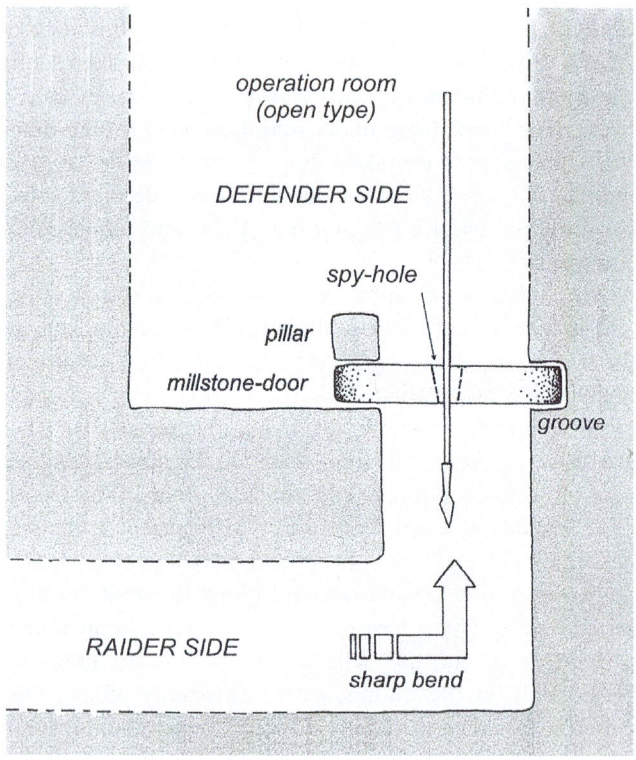

Fig. 3.14 Basic plan of an operation room (Bixio, 2012)

a passage referred to as the "vertical runner," which was slightly larger than the door's thickness and height, allowing it to roll securely into place. A corresponding groove on the opposite wall provided additional reinforcement against external pressure.

Basically it is the space between the support pillar or slab and the wall of the operation room, which houses the millstone door in an open position. It also serves as the runner to roll the door toward the groove. This space is slightly larger than the thickness of the millstone door and slightly higher than its diameter. There are horizontal runners too, with lowered trenches and/or raised parapets.

When the millstone door is in a closed position, the section of the tunnel situated opposite the vertical runner, in which the door is located, is referred to as the "groove." It is a vertical hollow excavated, normally with the same curve of the millstone, in the tunnel's wall opposite the vertical runner. It needs to block the millstone door while in closed position, avoiding its overturning in case of attackers' push, together with a second point of support consisting of the pillar or the slab.

In some operation rooms, a second vertical runner with a pillar on the opposite wall is used in place of a groove. This allows for the pushing of the millstone on both sides, rather than just one, thereby facilitating and accelerating the handling process when reopening. The maneuvering rooms with two pillars are a variation similar, in its conception, to the support provided by two slabs. This arrangement could be an innovation implemented at a later time because it has the advantage, compared with the single support with groove, of allowing the pushing of the millstone on both sides (push posts), instead of just one, easing and accelerating the handling, especially when reopening.

The wedge is the smallest but most vital component of an operational room. Millstone doors were blocked in either the closed or open position by a simple stone used as a wedge.

Operation rooms can come in many different designs. As well as different architectural styles such as open, enclosed, or hanging, these structures are combined with other defensive devices such as sharp bends, traps, and loopholes.

Other blockade systems may be found, such as barriers and shield doors—i.e., stone slabs inlaid in frames along a tunnel or on top of shafts—but they are not as common as millstone doors.

3.4.3 Barriers and Shield Doors

The identification of opposed vertical slits and slides in certain tunnels suggests the existence of a rudimentary blockade system. This system comprised round wooden trunks, which, during an attack, were stacked to create a defensive barrier. While the wooden components have been lost over time, the positioning of the slides indicates the direction of defense. In Cappadocia, only two such barrier housings have been discovered to date, located within the underground shelters at Karşı Kilise and Sivasa (Triolet & Triolet, 1993).

Shield doors were simple stone slabs, either square or rectangular, fitted into specially carved frames around tunnel or shaft openings. While the presence of these housing frames is common, intact slabs are rare. One such example has been documented at Filiktepe, measuring approximately 50 cm per side and 8 cm in thickness. These slabs featured a central hole, likely serving a dual function as both a spyhole and a handling mechanism. Some doors may have been wooden hatches, which, if they existed, were either lost or repurposed. These wooden barriers could have incorporated an advanced locking mechanism that was operable from the outside.

3.4.4 Tunnels

The primary defense mechanism relied on the narrow dimensions of tunnels, which forced intruders to advance in single file. This formation severely limited the ability of those behind to assist the person in front. Furthermore, in emergency escape routes, tunnel height was intentionally reduced, requiring movement on all fours (Fig. 3.15). If attackers located these passages, their hands would be occupied not only with carrying torches to illuminate the dark tunnels but also with maintaining balance. Consequently, they could only wield small weapons with restricted maneuverability. Retreat was difficult, and if the leading attacker was incapacitated, they would obstruct those following behind. Additionally, the use of battering rams was hindered by the deliberate placement of sharp turns just before doors, preventing direct forceful entry.

Fig. 3.15 Millstone door at the end of a tight tunnel at Büyük Büringüz underground shelter–Kayseri (photo A.E. Keskin)

3.4.5 Traps

Certain defensive mechanisms, referred to as "traps," are less conspicuous but highly effective. These traps are categorized into two types: horizontal "bottleneck" traps and vertical "shaft" traps. While they primarily served to slow down attackers, their true function was to injure, repel, or eliminate intruders. These traps were often used in conjunction with millstone doors, enhancing their defensive effectiveness.

3.4.5.1 Horizontal Traps

Horizontal traps involved an abrupt narrowing of tunnels in both height and width, compelling assailants to crawl through for a short distance. This position left them vulnerable as they emerged, with their heads lowered and arms engaged in supporting their movement and holding a light source. A hidden sentry, stationed in a wider section of the tunnel, could easily strike the attacker's head upon emergence. The fallen body would obstruct or slow down those following, discouraging further advances.

3.4.5.2 Vertical Traps

Vertical traps included both ascending and descending varieties. Ascending traps ranged from short climbs to tall vertical shafts. In the case of short ascents, attackers could mistake them for cul-de-sacs and hesitate. Should the intruder persist in his advance, he would inevitably become aware of the opening above him at the moment of impact. This would allow the defender to neutralize him with spears, arrows, or even a simple club, without the intruder being able to resist or fight back due to the narrowness of the passageway, which severely limited his movements (Fig. 3.16).

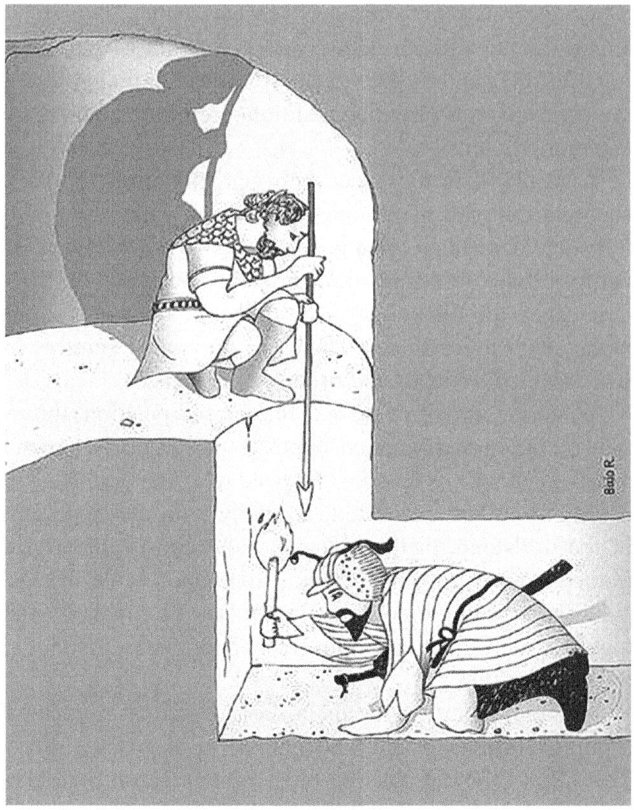

Fig. 3.16 An ascending trap (Bixio, 2012)

Obviously, although more rarely, this system should work in reverse, so from up to down, as it seems for the small shelter of Ağa Dam, near Ovaören, which has a double trap, in sequence, first upward then downward (Bixio, 2012).

In Cappadocia, ascending shaft traps whose entrance opens on the ceiling in a room instead of along a tunnel are frequent. In this case the lower entrance of the shaft would be probably reached with a portable ladder that was retired by defenders after they climbed into the upper room (redoubt) by means of footholds carved in the upper part of the pipe. Even more likely, simple ropes or rope ladders were used, tied at the top with rock rings or horizontal bars, let dangle, ready to use, and pulled back in the redoubt only in case of attack, after they went up.

The assailants, while preparing a system to climb up, would find themselves under the vertical of the shaft, exposed to the launch of projectiles which could consist of simple stones expressly stored previously in the redoubt. Even if the attackers would start to climb up, they should advance chimneying through the shaft with both hands occupied to hang on, therefore always at the mercy of the defenders.

Many of these shafts feature carved recesses, suggesting the presence of stone or wooden doors at their upper exits (Figs. 3.17 and 3.18).

Descending traps, on the other hand, were deep shafts positioned at tunnel entrances or within tunnels themselves. Some, such as those at Golgoli Shelter, also functioned as water wells. In shelters like Çomaklı in Kayseri Province, defenders could cover these shafts with boards, rendering them nearly impassable when removed during an attack. Similar strategies were observed in Vietnam, where thin, fragile coverings were used to deceive and trap intruders.

3.4.6 Modules and Systems

Large underground shelters were typically constructed by connecting smaller modular units over time. Each basic module, designed for one or several households, generally included the following components: Transit tunnel,

Fig. 3.17 A shaft/trap with a depth of 5 m in an underground shelter in Ürgüp (photo B. Yazlık)

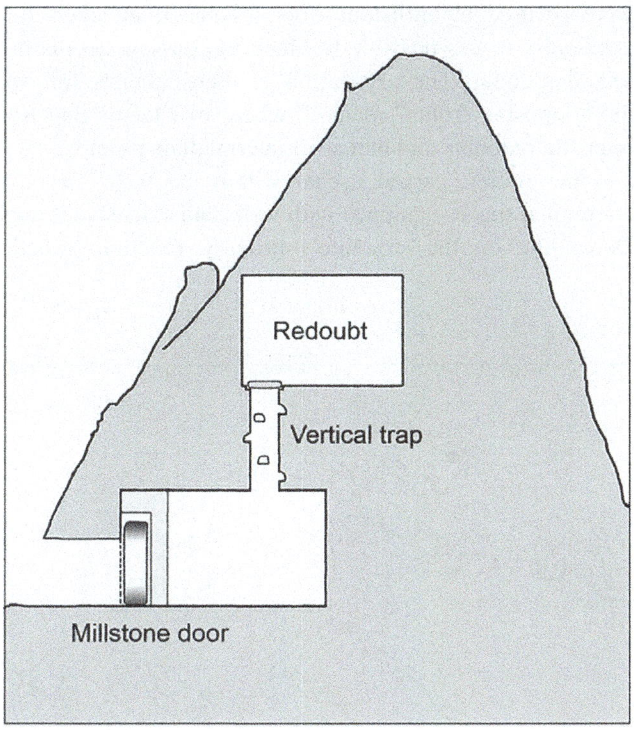

Fig. 3.18 A simplified sketch showing a vertical shelter and its shaft/trap (Bixio, 2012)

blockade mechanism, redoubt, and sometimes escape route way (Fig. 3.19). Larger systems also have connecting tunnels and inner rooms. By analyzing the shelters' plans, it appears evident that a kind of base module exists, schematically organized as follows:

Outer Belt Room: A space excavated within the rock part nearest the surface. It is typically large in size and is used for habitation or as a stable.

Transit Tunnel: A short tunnel that penetrates the interior of the rock. It is perpendicular to the cliff face and connects with an inner room (redoubt). It is approximately 180 cm high and between 80 and 90 cm wide. However, it is possible to find smaller sizes as well. This facilitates unobstructed passage for both people and animals.

Blockade mechanism/device: It consists of an operating room, situated centrally within the tunnel, equipped with a millstone door that can only be operated from within to impede access by potential intruders.

Redoubt Room: This chamber is located in the innermost belt of rock, providing a secure refuge for the inhabitants in the event of enemy raids. The room is equipped with warehouses or silos for food reserves, mangers, and, typically, a well that, being out of reach for the attackers, ensured the survival of besieged individuals for a sufficient duration. In some cases, a secondary blockade device is present at the debouchment of the connecting tunnel, also at a short distance from the first one, to augment the level of protection.

Escape Tunnel: Typically, a second tunnel originates from the redoubt, leading to another exit point on the cliffside or to another redoubt. This tunnel could be utilized to facilitate a hasty retreat in the event that the redoubt was about to be conquered. It could also be used to launch a surprise attack from behind the attackers or to receive or dispatch assistance to other groups of refugees. The dimensions of this tunnel, intended for emergency

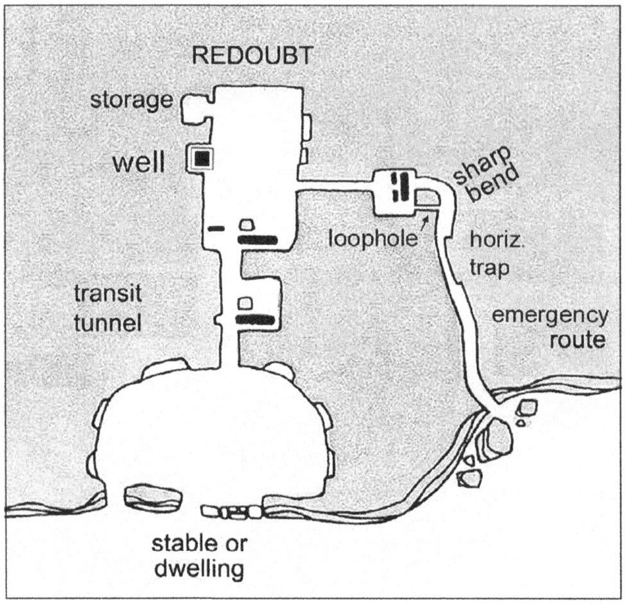

Fig. 3.19 Sketch of a module (Bixio, 2012)

situations, are notably constrained, with a width ranging from 60 to 70 cm and a height ranging from 60 to 100 cm. This restriction, necessitated by the need to traverse on all fours, is further augmented by the installation of a blockade mechanism. In addition to millstone doors, other defensive mechanisms may be present, such as sharp bends, loopholes, and traps, the purpose of which is to impede or delay access in the event that attackers have identified the entrance.

These modules were likely designed to provide refuge for a limited number of individuals, perhaps accommodating a single family unit. The redoubts are found also inside more articulated rock-cut residential or monastic complexes.

Much more larger and articulated underground shelters have been documented. They could house hundreds of people and obviously were intended for communities composed of several family units. It is likely to be expected that each familial unit used an independent module. However, every unit was not isolated but connected, with a predetermined pattern, through tunnels, to other adjacent modules so as to constitute a network excavated within the inner rock mass. Thus, a group of refugees under attack could receive help from other groups without exit from the underground structure, or could retreat to other sectors, in their turn properly secured, in case their shelter was about to fall into the hands of attackers. Therefore, all the modules, each made up of the elements described above and combined with other mechanisms/devices, constituted a system (Fig. 3.20).

Connecting tunnels, typically excavated within the rock's inner belt, linked different redoubts. These tunnels, longer than transit tunnels, ran parallel to the cliff face and were secured by millstone doors. Thus, if attackers took a redoubt, it was possible to block the passage to the next linked redoubt. The excavation of these tunnels followed the "opposing fronts" method, where two teams dug from opposite redoubts and met at an intermediate point.

Inner rooms, carved deeper within the rock, served as communal spaces equipped with wells and extensive storage. Depending on the structure's primary function—whether

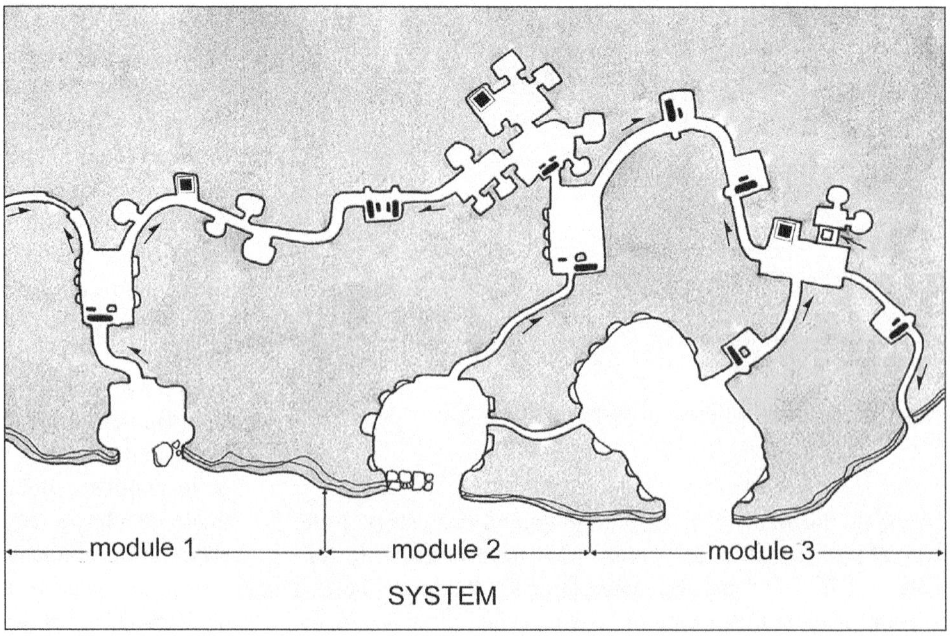

Fig. 3.20 Sketch of a system formed by the combination of three different modules (Bixio, 2012)

defensive or residential—some areas were reinforced with additional redoubts or multi-level rooms.

Inner rooms were additional rooms excavated within the innermost belt, protected by redoubts. These were intended to provide spaces more comfortable for the community, with wells and a large number of warehouses for food. Of course each complex has its own peculiarities so, depending on the case, it can have a more defensive or more residential vocation. Sometimes, some sections of the shelter are integrated with rooms excavated in upper or lower levels, with the same function of inner rooms or as additional redoubts.

In some cases, connecting or transit tunnels are protected with bottleneck horizontal traps or vertical traps, in place of or in addition to blockade devices.

3.4.7 Isolating Rooms and Isolated Areas

More specifically analyzing the reciprocal position of modules, their links through both types of tunnels (transit and connecting), the location and the defense direction of blockade devices, it seems that, in most cases, the excavation operations of various systems were not personal initiatives, but followed a carefully pre-specified plan. Indeed, every underground network shows a progressive defense organization designed so that one could block the attack from every direction of penetration by intruders and that, at the same time, it would ensure the stay and enough free movements for refugees inside the networks themselves.

Every network consists of a series of isolating rooms and isolated areas (Triolet & Triolet, 2002). The isolating rooms are in a central position and are provided with several entrances giving access, through connecting tunnels, to opposite sectors of the underground shelters. Every opening is protected by one or more blockade devices that could block the assailants from any direction, allowing open access to other sectors. Thus, a sector about to fall in raiders' hands could be abandoned, by isolating the other zones where the defenders could retire, to reorganize the defense. This method, in largest shelters, could repeat many times.

In consideration of the defensive strategy employed by the installation of blockade devices, and the potential direction of penetration by intruders, it is possible to discern the existence of several defensive positions, organized in a progressive sequence. These include defensive positions that are mutually supportive, as well as those that are oriented toward defending against attacks from opposing directions. This defensive configuration allows for the insulation of the innermost sections of the installation from attacks originating from both the current access point and from other tunnels that have been established communication with contiguous sectors. In addition, it provides protection from attacks that may originate from other currently inaccessible entrances. It is evident that, based on the defensive orientation of each blockade mechanism, the tunnel or the room on the defenders' side could also have a dead-end, devoid of alternative routes. Conversely, on the attackers' side, an external access point, either direct or via alternative pathways, must exist, even if it is currently obstructed. This observation can assist in comprehending the defensive rationale behind the structure and potentially inform the design of subsequent investigations.

3.5 Excavation Techniques of Underground Shelters

The excavation techniques to achieve underground structures are an interesting topic that can be further detailed. One can notice marks of many kinds of interventions depending on the type of structure to achieve. Understanding the nature and archaeology of rock-cut architecture necessitates an understanding of its substance and the process of carving.

The traditional method of carving rock-cut spaces is through the consecutive removal of large stone blocks. Based on Öztürk's experimental studies, we know that under the best conditions, a worker can excavate up to 1 m^3/day (Öztürk, 2009, 2017). Unforeseen characteristics in the rock mass often forced carvers to improvise, even changing the direction or dimensions of a space during the carving process. Traditionally, the stone block is detached from the main rock by carving a deep channel around it, using pickaxes of different weight and length. After breaking the surrounding rock mass, wedges are inserted into the channel along one side and hit with a sledgehammer until the block breaks away.

Here we list two particular excavation techniques used to achieve tunnels and water works: the excavation on opposing fronts and the use of pilot holes.

3.5.1 Excavation on Opposing Fronts

The excavation on opposing fronts is a widespread age-old technique to achieve underground tunnels. It has been largely documented in the water tunnels in Greece and in Italy, or in the qanat/karez in China, Persia and all the Sub-Saharan area in Africa (Castellani, 2001; Wessels, 2000), as well as in many places in Asia Minor.

For example, it is very clear that the 1,637 m long water tunnel surveyed in Gaziantep was dug from opposing fronts. In this tunnel, which has a total of 24 surface connections, many junction errors were detected both horizontally and vertically (Yamaç & Okuducu, 2017).

In Cappadocia, considering the large number of structures excavated within rocks and underground, marks of this technique are virtually everywhere. This evidence usually permits to develop reliable assumptions about the building sequence of settlements, especially for underground shelters and water works. Moreover, on several occasions the interpretation of excavation directions helps to locate the entry points to new sectors for some surveyed cavities.

Güvercinlik Valley Tunnel in Uçhisar is the longest water tunnel ever measured in Cappadocia. With a total length of 3,600 m, 51 separate surface connections and several side tunnels, that water channel is a marvelous example of hydraulic engineering of Middle Ages (Gilli & Yamaç, 2015). Just like all other tunnels, it is not possible to excavate such a long tunnel from a single direction. Moreover, the junction errors encountered in all other tunnels are abundantly present in Güvercinlik Valley Tunnel, which is evidence of opposing fronts excavation (Fig. 3.21).

Essentially, the opposing fronts technique consists of an underground excavation, starting from different independent accesses, of two tunnels, one toward the other, in order to join them as a single one or as an intercommunicating network by reducing the working time.

The direction of fronts' progress is recognized thanks to the marks of excavation tools left on the side walls of the tunnels, probably a pointy stone hammer or a pick with a short handle, very likely similar to those still in use today. Marks left by these tools are caused by the forward movement of the carver's arm hitting the rock, with arched-shape trajectories or with a complete movement with a semicircle shape, where the arrow indicates the direction of progress (Fig. 3.22). Moreover, the collision normally leaves on the rock a shallow mark on the starting point and progressively deepest one at the end, changing the angle depending on where it falls.

It is believed that the work was carried out by two teams at the same time, that starting from two independent entrances, or origin points, would meet in an underground area approximately prearranged, depending on the excavation rate and the rock features, defining the progress direction of the tunnel thanks to some expedients: alignment shafts, aligning light, noise, and vibrations propagation.

It should be noted that, in some cases, it is assumed that the origin points are as many as the excavating teams of each segment, drastically reducing the timeline. To illustrate, a tunnel measuring 170 m in length, 1.5 m in height, and 80 cm in width equates to 200 cubic meters of excavation. According to the calculation of one cubic meter excavated per day per individual, it would require one individual 200 days to complete this excavation, whereas a team of eight individuals can accomplish the same task in just 25 days (Bixio et al., 2023).

The tunnels excavated with opposing fronts were aligned and then connected by listening to the noise and vibrations propagation, through the rock mass, made by the operators on both ends. It would not be surprising if they had also adopted other measures documented in ancient sources which describe criteria to excavate the so-called quarry and counter-quarry tunnels, during sieges of walled cities. Indeed, the techniques utilized during wars to discover enemies' underground activities and to advance the fronts of

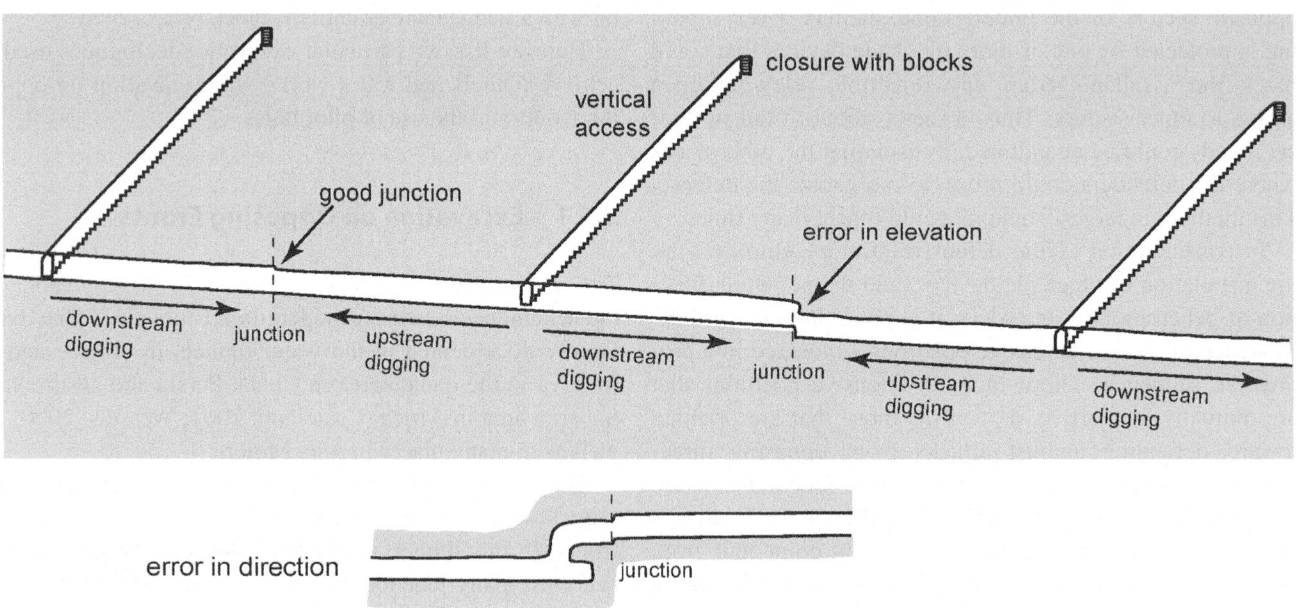

Fig. 3.21 Typical opposing fronts excavation and errors in Güvercinlik water tunnel (drawn E. Gilli) (Gilli & Yamaç, 2015)

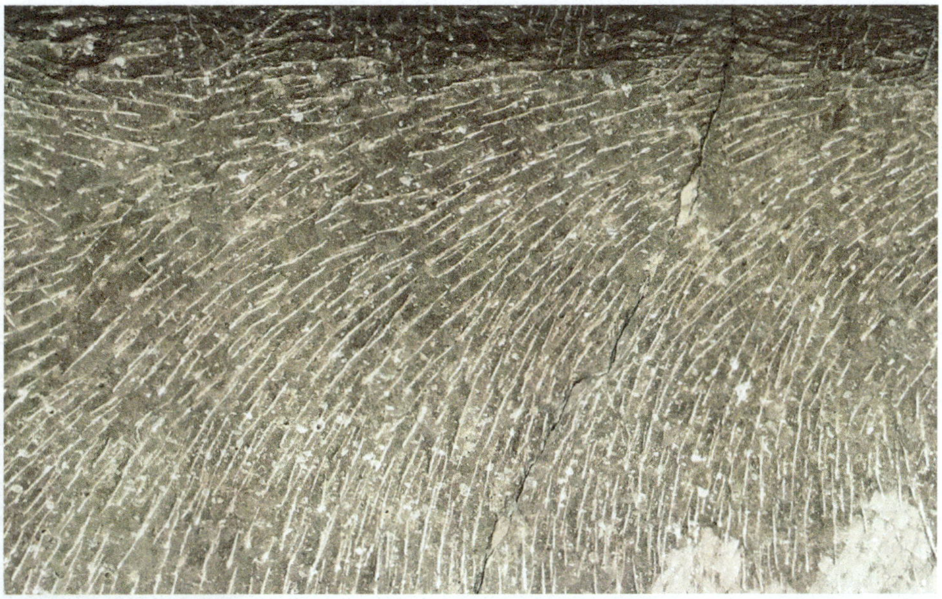

Fig. 3.22 Semicircular axe marks on the wall of a shelter in Ürgüp (photo A. Yamaç)

the tunnels toward the prefixed target are many and ingenious, all based on the propagation of vibrations through the rock of pickaxe strikes of miners, rather than on the perception of the noise of excavation.

3.5.2 Excavation Phases

If the elevation between the surface and the anticipated level of the subterranean tunnel is minimal, it would be advantageous to create several alignment shafts. This approach would allow for the simultaneous excavation of tunnels from their bases, extending in both directions toward the adjacent shafts (Fig. 3.23).

We shall designate the term "divergence point" to refer to the origin of fronts that are advancing in opposing directions, while "junction point" will signify the location where these fronts converge and intersect.

In instances where the depth between the surface and the tunnel was excessive, the technique known as coltellatio was employed, which involved the excavation of a single shaft, or at most two, positioned near both entrances. The process involved the careful alignment and leveling of several ranging poles on the terrain, keeping to the natural morphology, in order to ascertain the altitude and the entry directions at both ends of the tunnel to be constructed. Subsequently, one or two shafts on either side were carefully excavated, in alignment with the previously mentioned ranging poles. The shafts functioned as a means of alignment for the tunnel, which was concurrently excavated from both ends.

In Cappadocia, during the excavation of tunnels, typically characterized by a limited extent, the alignment shafts were not employed, leading to a "blind excavation" approach. This method resulted in unpredictable trajectories, marked by more significant directional shifts.

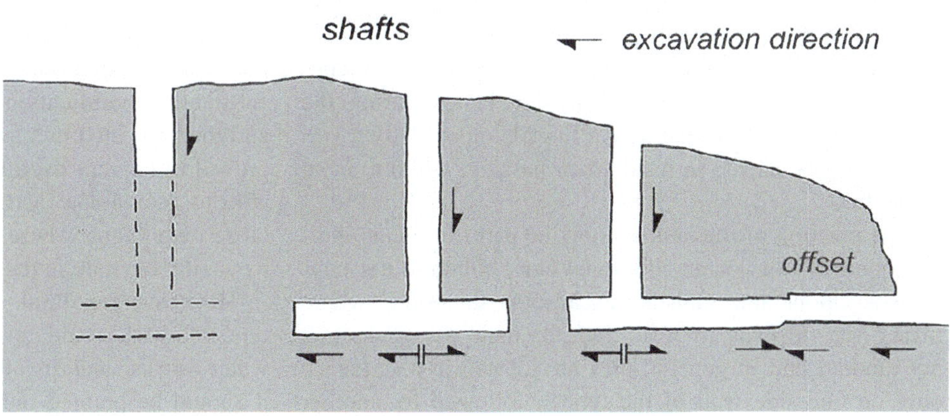

Fig. 3.23 Tunnel excavation on opposing fronts with shafts (Bixio, 2012)

The error further decreased using the "winding excavation" system between the first two shafts. Thus, the light from the entrance was accurately aligned. Then it was sufficient to continue to excavate without losing sight of the rays of a lamp's light. In this case, the error could decrease to less than one meter.

While achieving opposite tunnels the critical point was their junction. Once they reached a distance between fronts adequate for the excavation noise propagation within the rock mass, one of two tunnels, or both, deviated, also repeatedly, toward the source of sound waves and/or vibrations. The procedure could occur just meters away from the junction point; otherwise, the first adjustments could start more than 20 m before and then definitively be corrected in the last two meters.

For all these reasons, normally, the meeting point of opposing fronts, characterized by the inversion of excavation marks, corresponds to a change of direction of tunnels, more or less abrupt depending on the error margin.

It therefore follows that, at the right time, the presence of one person would be necessary on the opposing front to produce the noise source needed to direct right to the excavation. Therefore, in the event of a junction point, we know for sure that at least a second entrance to the underground structure exists or existed, even if a collapse or a filling nowadays closes the passage.

The survey of excavation marks, partially combined with the positioning of niches for lamps and, for shelters, with the defense direction of millstone doors, allows to properly evaluate the building logic and the evolution of underground structures respected by the designers-users.

3.5.3 Niches for Lamps

Another recurring element within tunnels are the small niches achieved to arrange lamps so as to illuminate the excavated area and the path for the debris removal operators. They are scattered along the path, normally near the ceiling springer and the distance from each other probably corresponds to the daily rate of the front progress or, anyway, where they were needed.

After the excavation, within the water works, niches could still be used to temporarily lay lamps during the maintenance, while in the transit tunnels most likely a single lamp was carried in hand when passing, without using the niches.

The position of the lamps along the path can be an indirect confirmation, or an additional clue, although not conclusive, about the tunnels' excavation direction, where marks are difficult to read. Indeed, their distribution is not random but, in general, they are located in a sequence only on one side wall of the tunnel, followed by another sequence on the opposite side. This change of side usually corresponds to the divergence point, or to the junction point, so to the reversal of work fronts.

Fig. 3.24 An antique oil lamp in a rock-cut niche (photo A. Yamaç)

Moreover, from the surveys carried out, it seems that lamps are usually located on the left side in relation to the excavation direction progress; this arrangement corresponds to the best lighting position for a carver who excavates with his right arm (Fig. 3.24).

3.5.4 Pilot Holes

Pilot holes consist of small perforations of the rock to verify, and eventually correct the alignment and the distance of the opposing fronts in excavation works, so as to preserve the dividing rock screen, minimizing the thickness, without breaking it down. Probably, they were achieved with tools like drill rods with a diameter of around 8/10 cm. From a few examples documented until now we note that, in case of error, if just a limited correction was needed, the pilot hole was repeated with a little variance, whereas in case of substantial modification of the excavation it was sheared and then repeated in a better suited position. This technique proved to be very useful to build underground basins, suitable to accumulate and store water, creating a barrier, or close diaphragm, where the catchment devices are located, by saving the rock in the excavation.

However, we think that more investigations on this type of work could reveal an extensive use, not just for water works, and in other areas than Göreme. However, it should be pointed out that, in case pilot holes served to align opposite tunnels, then connected by eliminating the diaphragm, probably, in most cases every mark would have disappeared, making detection impossible.

3.6 List of Underground Shelters

It is nearly impossible to compile a comprehensive and accurate list of the underground shelters in Cappadocia. The initial and perhaps most significant reason for this is the absence of an inventory that comprehensively identifies all of these structures on the site.

While the three databases currently available encompass on-site detection and surveys, some shelters have been taken from different sources, and their confirmation is uncertain.

Additionally, the fact that the same place sometimes goes by different names complicates the situation. Furthermore, the discovery of additional underground shelters even during the preparation of this book illustrates the dynamic nature of such a list. It is reasonable to assume that numerous other shelters remain undiscovered, particularly in the vast region spanning 50,000 km^2 and comprising numerous villages.

The territorial distribution of underground shelters in the Cappadocia region gives an idea about the regional distribution of this architectural tradition. From the point of view of territorial distribution, considering as identifying marks of underground shelters the presence of millstone doors or their operation rooms, which we know for certain, we can consider the limits hereafter listed.

–Westernmost shelter of the region is Çıralı Sinkhole, located in the area named Obruk Yaylası, 80 km southwest of Aksaray, in Konya Province;
–Easternmost site is Pazarören, 60 km east of Kayseri. The distance from Çıralı Sinkhole is about 270 km;
–Northernmost shelter is Çeşka, 3 km northeast of Yozgat, main town of the same province;
–Southernmost shelter is located at the village of Fertek, about 3 km west of Niğde, main town of the same province, about 200 km south of Çeşka.

If we accept these four shelters as the corners of a rectangle, we see a region close to Cappadocia Province in Fig. 2.1 and the area of this rectangle is approximately 55,000 km^2 but the region where the shelters are densely located is about 20,000 km^2 (Figs. 3.25 and 3.26).

The list below includes all underground shelters that have been registered so far and have at least one reference. Of these 253 shelters, 90 have been surveyed. In developing this list of underground shelters, several criteria were considered, as outlined below:

- The list includes only the underground shelters in the six provinces of today, namely Aksaray, Nevşehir, Kayseri, Kırşehir, Yozgat, and Niğde, which comprise ancient Cappadocia. Although there are few shelters outside these provinces, they have been excluded.
- Three main sources were used in the preparation of this list: *Cappadocia: Records of the underground sites* (Bixio, 2012), *Türkiye Arkeolojik Yerleşmeleri 8—Bizans/Iç Anadolu* (Akyürek et al., 2015), *Artificial Cavities of Turkey; Updating the Map of Anthropogenic Cavities in the Mediterranean Basin* (Bixio et al., 2021).
- In making the following list from these three sources, shelters whose references are questionable or have not been confirmed by a second source are highlighted with a question mark "?" in the last column. Also, the "X" mark in the same column indicates underground shelters that have been scientifically researched and surveyed to some extent.

Name	Province	Distr	Village	
Acıgöl 1	Nevşehir	Nevşehir	Acıgöl	X
Acıgöl 2 (İğdeli)	Nevşehir	Nevşehir	İğdeli	
Acıgöl 3 (Topada)	Nevşehir	Nevşehir	Ağıllı	X
Acıpınar	Aksaray	Aksaray	Acıpınar	
Ağadam	Nevşehir	Gülşehir	Ovaören	X
Ağırnas	Kayseri	Melikgazi	Ağırnas	X
Akdam (Aktepe)	Nevşehir	Gülşehir	Gülpınar	X
Aktaş	Niğde	Niğde	Aktaş	
Akmezar 1	Aksaray	Gülağaç	Akmezar	
Akmezar 2 (Hıcıp)	Aksaray	Gülağaç	Akmezar	
Aksaklı 1	Kırşehir	Mucur	Aksaklı	X
Aksaklı 2	Kırşehir	Mucur	Aksaklı	X
Alanyurt	Aksaray	Güzelyurt	Alanyurt	
Alayhanı	Aksaray	Gülağaç	Alayhanı	
Ali Dağı	Kayseri	Talas	Talas	X
Ali Saip Paşa	Kayseri	Talas	Talas	X
Asmakaradam	Kırşehir	Mucur	Asmakara	X
Altınyazı 1	Kırşehir	Mucur	Altınyazı	X
Altınyazı 2	Kırşehir	Mucur	Altınyazı	X
Altınyazı 3	Kırşehir	Mucur	Altınyazı	X
Aşağı Hasinli	Yozgat	Boğazlıyan	Aş. Hasinli	X
Avanos 1	Nevşehir	Avanos	Avanos	X
Avanos 2	Nevşehir	Avanos	Avanos	X
Ayan (Derinkuyu 5)	Nevşehir	Derinkuyu	Derinkuyu	X

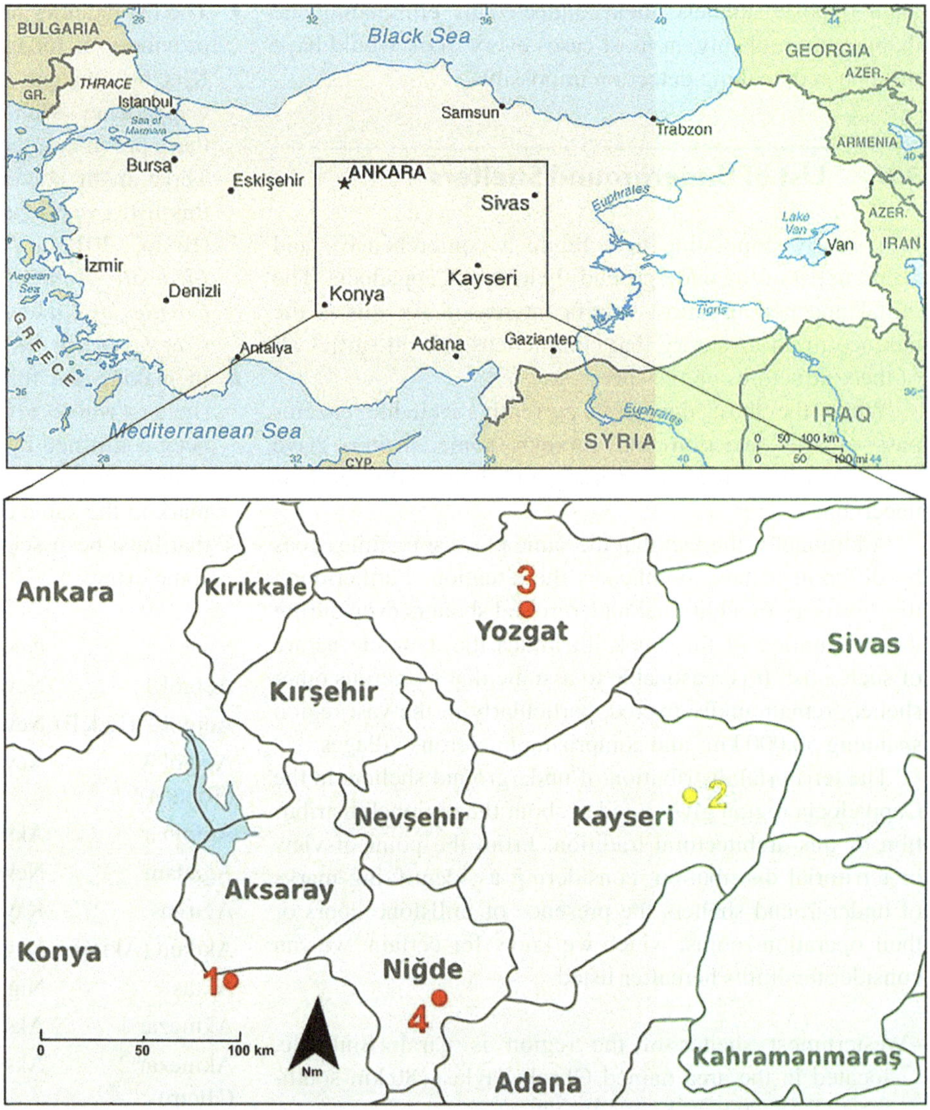

Fig. 3.25 Underground shelters at the farthest points of Cappadocia; 1. Çıralı 2. Pazarören 3. Çeşka 4. Fertek (after Google Maps—elaboration A. Yamaç)

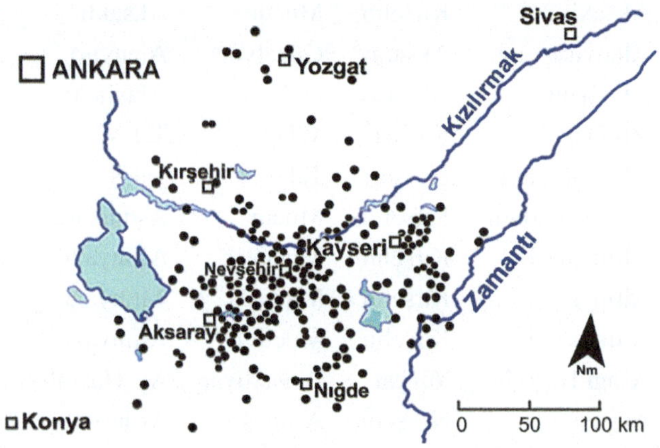

Fig. 3.26 Underground shelter density map (drawn A. Yamaç)

Name	Province	Distr	Village	
Aynalı Church	Nevşehir	Nevşehir	Göreme	X
Azıklı	Nevşehir	Nevşehir	Azıklı	
Bağlıca	Nevşehir	Acıgöl	Bağlıca	
Bağlıkaya	Aksaray	Aksaray	Bağlıkaya	
Bahadın	Yozgat	Sorgun	Bahadın	
Balak Ramazan	Aksaray	Güzelyurt	Manastır	
Başaran	Aksaray	Eskil	Başaran	?
Bebek	Aksaray	Aksaray	Bebek	
Belağası 1	Kayseri	Gesi	Belağası	X
Belağası 2	Kayseri	Gesi	Belağası	X
Belha Manastırı	Nevşehir	Avanos	Özkonak	X
Belisırma	Aksaray	Güzelyurt	Belisırma	
Bezirhane	Kırşehir	Mucur	Asmakara	
Boğazköy	Nevşehir	Nevşehir	Boğazköy	X

Name	Province	Distr	Village	
Bozcatepe (Lefkere)	Aksaray	Aksaray	Bozcatepe	
Bozcayurt (Mandama)	Aksaray	Aksaray	Bozcayurt	
Bölükören	Nevşehir	Gülşehir	Bölükören	
Bünyan	Kayseri	Bünyan	Bünyan	X
Büyük Bürüngüz	Kayseri	Melikgazi	B. Bürüngüz	X
Büyükgüve	Aksaray	Aksaray	Büyükgüve	
Canavar Monastery	Kayseri	Yeşilhisar	Soğanlı	X
Çağıl	Aksaray	Aksaray	Çağıl	
Çakıllı	Nevşehir	Derinkuyu	Çakıllı	
Çanlı Kilise	Aksaray	Aksaray	Akhisar	X
Çardak 1	Aksaray	Aksaray	Altınkaya	
Çardak	Nevşehir	Nevşehir	Çardak	
Çatalin	Kayseri	Gesi	Kayabağ	X
Çatalsu	Aksaray	Gülağaç	Çatalsu	
Çayırlı	Niğde	Niğde	Çayırlı	
Çekiçler	Aksaray	Aksaray	Çekiçler	
Çeltek	Aksaray	Aksaray	Çeltek	
Çerkezbayın	Aksaray	Güzelyurt	Selime	
Çeşka	Yozgat	Yozgat	Çeşka	
Çıralı Sinkhole	Konya	Karapınar	Karapınar	X
Çiftlikköy	Nevşehir	Nevşehir	Çiftlikköy	
Çolaknebi	Aksaray	Aksaray	Çolaknebi	?
Çomaklı	Kayseri	Develi	Çomaklı	X
Çömlekçi	Aksaray	Güzelyurt	Çömlekçi	
Çukurören	Aksaray	Gülağaç	Camiliören	
Danalık	Kayseri	Talas	Çömlekçi	X
Değirmenli	Niğde	Niğde	Değirmen	X
Derinkuyu	Nevşehir	Derinkuyu	Derinkuyu	X
Dikilitaş	Niğde	Niğde	Dikilitaş	
Doğala	Nevşehir	Derinkuyu	Doğala	X
Doğanlı	Kayseri	Yeşilhisar	Doğanlı	X
Doğantarla	Aksaray	Aksaray	Doğantarla	
Dorukini	Aksaray	Aksaray	Dorukini	
Durmuş (Derinkuyu 4)	Nevşehir	Derinkuyu	Derinkuyu	X
Düğüz	Aksaray	Gülağaç	Düğüz	
Edikli	Niğde	Niğde	Edikli	
Elmacık (Gine)	Aksaray	Aksaray	Elmacık	
Emiruşağı	Kayseri	Tomarza	Emiruşağı	X
Erdoğdu	Aksaray	Eskil	Eşmekaya	

Name	Province	Distr	Village	
Eski Pusatlı	Kayseri	Tomarza	Pusatlı	X
Eskil	Aksaray	Eskil	Eskil	
Fakılı 1 (Damlalı)	Yozgat	Yenifakılı	Yenifakılı	
Fakılı 2	Yozgat	Yenifakılı	Yenifakılı	
Fatinler	Kayseri	Tomarza	Şıh Barak	X
Fertek	Niğde	Niğde	Fertek	
Fırıntaş	Aksaray	Güzelyurt	Manastır	
Fırkatan (Aynalı Kilise)	Nevşehir	Nevşehir	Göreme	
Filiktepe (Filikören)	Nevşehir	Acıgöl	Ovaören	X
Gaziemir	Aksaray	Güzelyurt	Gaziemir	
Gelegüle	Aksaray	Aksaray	Sevinçli	
Gelesin (Babakonağı)	Aksaray	Aksaray	Gelesin	
Germir	Kayseri	Melikgazi	Germir	?
Gidiriç (Kepez)	Aksaray	Aksaray	Akçakent	
Gomeda	Nevşehir	Ürgüp	Mustafapaşa	X
Golgoli 1	Nevşehir	Ürgüp	Cemil	X
Golgoli 2	Nevşehir	Ürgüp	Cemil	X
Gökçe	Aksaray	Aksaray	Gökçe	
Gökçe 2 (Viranşehir)	Aksaray	Aksaray	Gökçe	
Göktaş	Aksaray	Gülağaç	Demirci	
Gölcükbaşı	Niğde	Niğde	Yeşilgölcük	
Göre	Nevşehir	Nevşehir	Göre	
Göreme Valley	Nevşehir	Nevşehir	Göreme	
Göstesin (Ovaören)	Nevşehir	Gülşehir	Ovaören	X
Göynük	Nevşehir	Avanos	Göynük	
Gözlükuyu (Munamak)	Aksaray	Aksaray	Gözlükuyu	
Gözkayası	Aksaray	Güzelyurt	Gözkayası	
Gülbayır	Kayseri	Yeşilhisar	Gülbayır	?
Gülistan	Kayseri	Talas	Gülistan	X
Gülpınar (Hıcıp)	Aksaray	Gülağaç	Gülpınar	
Gümüşkent	Nevşehir	Gülşehir	Gümüşkent	?
Gürlek Dere	Nevşehir	Gülşehir	Yeşilöz	
Güzelöz	Kayseri	Yeşilhisar	Güzelöz	X
Güzelyurt (Gelveri)	Aksaray	Güzelyurt	Güzelyurt	X
Hacıgaybı	Aksaray	Aksaray	Yenipınar	
Harman Yeri	Kayseri	Talas	Talas	X
Hasanköy	Niğde	Niğde	Hasanköy	?

Name	Province	Distr	Village	
Helvadere	Aksaray	Aksaray	Helvadere	
Höbek	Kayseri	Kocasinan	Höbek	X
Ihlara 1 (Mumyalar)	Aksaray	Güzelyurt	Yeşilköy	
Ihlara 2 (Kale Yeraltı)	Aksaray	Güzelyurt	Yeşilköy	X
Ilısu	Aksaray	Güzelyurt	Ilısu	?
İçik	Nevşehir	Nevşehir	İçik	
İncesu 1	Aksaray	Aksaray	İncesu	
İncesu 2 (Andiklik)	Aksaray	Aksaray	İncesu	
İncesu 3 (Musa Kuyusu)	Aksaray	Aksaray	İncesu	
İncili	Kayseri	Tomarza	İncili	X
İnözü	Yozgat	Yozgat	Köseyusuflu	
Kadı Kalesi	Nevşehir	Ürgüp	Ürgüp	?
Kanlısivri S2	Nevşehir	Göreme	Göreme	X
Kanlısivri S10	Nevşehir	Göreme	Göreme	X
Karaağaç	Kayseri	Talas	Reşadiye	?
Karaasmaz	Niğde	Altunhisar	Altunhisar	?
Karaburna	Nevşehir	Hacibektaş	Karaburna	
Karacalı 1	Kırşehir	Mucur	Karacalı	X
Karacalı 2	Kırşehir	Mucur	Karacalı	X
Karacalı 3	Kırşehir	Mucur	Karacalı	X
Karacaören	Aksaray	Aksaray	Karacaören	
Karakaya	Nevşehir	Acıgöl	Tatlarin	
Karakoç	Yozgat	Boğazlıyan	Karakoç	
Karanlık Church	Nevşehir	Nevşehir	Göreme	X
Karataş 1 (Bektaşlı)	Aksaray	Aksaray	Karataş	
Karataş 2 (Zeyne)	Aksaray	Aksaray	Karataş	
Kargın Yaylası	Aksaray	Aksaray	Altınkaya	?
Karşı Kilise 1	Nevşehir	Gülşehir	Gülşehir	X
Karşı Kilise 2	Nevşehir	Gülşehir	Gülşehir	X
Kavlaktepe	Niğde	Çamardı	Kavlaktepe	
Kayabaşı	Kayseri	Bünyan	Bünyan	
Kayırlı	Niğde	Niğde	Kayırlı	
Kaymaklı	Nevşehir	Nevşehir	Kaymaklı	X
Kaz Kale	Nevşehir	Derinkuyu	Kuyulutatlar	
Kemer Ören (Kırkgöz)	Nevşehir	Gülşehir	Ovaören	X
Kepez	Kırşehir	Mucur	Kepez	X
Keşlik Monastery	Nevşehir	Ürgüp	Cemil	
Kılıçlar Church	Nevşehir	Nevşehir	Göreme	X
Kırkgöz	Aksaray	Aksaray	Saratlı	
Kırağcaşar	Niğde	Niğde	Ağcaşar	
Kırk Şehitler	Nevşehir	Ürgüp	Şahinefendi	X
Kırlangıç	Kayseri	Melikgazi	Hisarcık	X
Kızılkaya	Aksaray	Gülşehir	Kızılkaya	
Kızılören	Kayseri	Tomarza	Kızılören	
Kızılören Yaylası	Nevşehir	Derinkuyu	Derinkuyu	
Kirkor	Kayseri	Tomarza	Tomarza	X
Konaklı (Misli)	Niğde	Niğde	Konaklı	
Kömür	Kayseri	Tomarza	Kömür	X
Kulluk Ali	Aksaray	Güzelyurt	Manastır	
Kulpak	Kırşehir	Mucur	Kulpak	X
Kuruköprü	Kayseri	Talas	Kuruköprü	X
Kuşcağız	Kayseri	Tomarza	Kuşcağız	X
Kuzey Kalesi	Aksaray	Aksaray	Altınkaya	?
Küçükgüve	Aksaray	Aksaray	Küçükgüve	
Kürt Kasımlı	Nevşehir	Gülşehir	Abuşağı	
Kümbetaltı	Kırşehir	Kırşehir	Kırşehir	X
Macarlı	Aksaray	Aksaray	Macarlı	?
Mahmatlar	Nevşehir	Avanos	Mahmatlar	
Mahmutçuk	Kayseri	Develi	Yenice	X
Mamasun	Aksaray	Aksaray	Gökçe	?
Manastır (Derinkuyu 2)	Nevşehir	Derinkuyu	Derinkuyu	
Manastır Vadisi 1	Aksaray	Güzelyurt	Manastır	
Manastır Vadisi 2 (Kulluk)	Aksaray	Güzelyurt	Manastır	
Manastır Vadisi 3 (Balak)	Aksaray	Güzelyurt	Manastır	
Manastır Vadisi 4 (Fırınt.)	Aksaray	Güzelyurt	Manastır	
Mazı 1	Nevşehir	Ürgüp	Mazıköy	X
Mazı 2 (Cevriye)	Nevşehir	Ürgüp	Mazıköy	X
Mecemir	Aksaray	Aksaray	Koçpınar	
Melendiz Valley	Aksaray	Aksaray	Belisırma	X
Meskendir Valley	Nevşehir	Nevşehir	Göreme	X
Mimar Sinan	Kayseri	Melikgazi	Ağırnas	X
Mucur	Kırşehir	Mucur	Mucur	X
Musa Kuyusu	Aksaray	Aksaray	İncesu	
Mustafa Evi	Kayseri	Melikgazi	Ağırnas	X
Oymalı	Konya	Karapınar	Oymalı	
Ozancık	Aksaray	Ortaköy	Ozancık	

Name	Province	Distr	Village	
Ören	Nevşehir	Gülşehir	Yeşilöz	X
Örentepe	Nevşehir	Nevşehir	Kurugöl	X
Ötedere Vadisi No. 1	Kayseri	Gesi	Belağası	X
Ötedere Vadisi No. 2	Kayseri	Gesi	Belağası	X
Özkonak (Genezin)	Nevşehir	Avanos	Özkonak	X
Özler	Yozgat	Boğazlıyan	Özler	X
Pazarören	Kayseri	Pınarbaşı	Pazarören	X
Penzikli	Kayseri	Gesi	Kayabağ	X
Pınarca	Aksaray	Güzelyurt	Kumluk	
Pınarbaşı	Aksaray	Gülağaç	Pınarbaşı	
Pörnekler	Aksaray	Aksaray	B. Pörnek	
Saraycık Öreni	Aksaray	Aksaray	Altınkaya	?
Sarıağıl	Aksaray	Aksaray	Sarıağıl	
Selime	Aksaray	Güzelyurt	Selime	
Sağırkaraca	Aksaray	Aksaray	Sağırkaraca	
Sığırlı	Nevşehir	Gülşehir	Yeşilyurt	
Sivasa (Gökçetoprak)	Nevşehir	Gülşehir	Gökçetoprak	X
Sofular	Nevşehir	Ürgüp	Sofular	
Sofular	Aksaray	Gülağaç	Sofular	
Sosun	Kayseri	Talas	Sosun	X
St Eustachius 1	Nevşehir	Göreme	Göreme	X
St Eustachius 2	Nevşehir	Göreme	Göreme	X
St Eustachius 3	Nevşehir	Göreme	Göreme	X
St Eustachius 4	Nevşehir	Göreme	Göreme	X
St. Mercurius	Aksaray	Aksaray	Saratlı	X
Subaşı 1	Kayseri	Ağırnas	Subaşı	X
Subaşı 2	Kayseri	Ağırnas	Subaşı	X
Subaşı 3	Kayseri	Ağırnas	Subaşı	X
Susadı	Aksaray	Aksaray	Susadı	
Sümengen	Kayseri	Tomarza	Pusatlı	X
Suvermez	Nevşehir	Derinkuyu	Suvermez	
Şahinefendi 1 Orta Tepe	Nevşehir	Ürgüp	Şahinefendi	X
Şahinefendi 2 Yarlağan	Nevşehir	Ürgüp	Şahinefendi	X
Şahinefendi 3 Düğen Tepe	Nevşehir	Ürgüp	Şahinefendi	X
Şahmelik 1	Kayseri	Develi	Şahmelik	X
Şahmelik 2	Kayseri	Develi	Şahmelik	X
Şahmelik 3	Kayseri	Develi	Şahmelik	X
Şeyhler	Aksaray	Aksaray	Şeyhler	
Talasıra	Kayseri	Melikgazi	Güzelköy	X
Taşpınar	Aksaray	Aksaray	Taşpınar	
Tatlarin	Nevşehir	Acıgöl	Tatlarin	X
Tatlıca	Aksaray	Aksaray	Tatlıca	
Tekkaya	Nevşehir	Göreme	Çavuşin	X
Tıraz	Niğde	Çiftlik	Bozköy	
Tırhan	Niğde	Niğde	Tırhan	?
Tilköy	Nevşehir	Derinkuyu	Tilköy	
Tokarız (Dikmen)	Aksaray	Aksaray	Dikmen	?
Tombak	Kayseri	Develi	Tombak	X
Uzunkaya (Eskinuz)	Aksaray	Güzelyurt	Uzunkaya	
Vekse	Kayseri	Melikgazi	Vekse	
Yağan	Aksaray	Aksaray	Yağan	?
Yallı Damı	Nevşehir	Avanos	Özkonak	X
Yaprakhisar	Aksaray	Güzelyurt	Yaprakhisar	
Yenipınar (Hacıgaybı)	Aksaray	Aksaray	Yenipınar	
Yeniyuva (Nurgöz)	Aksaray	Aksaray	Nurgöz	
Yeşilöz (Tağar)	Nevşehir	Ürgüp	Yeşilöz	
Yeşilyurt	Niğde	Altunhisar	Yeşilyurt	
Yıldıztepe	Niğde	Niğde	Yıldıztepe	?
Yunak	Konya	Yunak	Karataş	
Zelve	Nevşehir	Avanos	Çavuşin	
Zemi Valley	Nevşehir	Nevşehir	Göreme	X
Zile (Özlüce)	Nevşehir	Derinkuyu	Özlüce	X
Zincidere	Kayseri	Talas	Zincidere	

References

Akok, M., & Gürçay, H. (1965). Yeraltı Şehirlerinde Bir İnceleme. *Türk Arkeoloji Dergisi, XIV*(1–2), 35–69.

Akyürek, E., Çömezoğlu, Ö., Tiryaki, A., Yamaç, A., Karakaya, N., Mimiroğlu, M., Evcim, S., et al. (2015). *Türkiye Arkeolojik Yerleşmeleri 8 - Bizans/Iç Anadolu.* Istanbul Ege Yayınları.

Ash, J. (2006). *A Byzantine journey* (2nd ed.). Tauris Parke Paperbacks.

Aydan, Ö., & Ulusay, R. (2013). Geomechanical evaluation of Derinkuyu antique underground city and its implications in geo-engineering. *Rock Mechanics and Rock Engineering, 46*, 731–754.

Ayhan, A. (2004). *Geological and morphological investigations of the underground cities of Cappadocia using GIS.* (PhD). ODTU, Ankara.

Bertucci, G., Bixio, R., & Traverso, M. (1995). *Le città sotterranee della Cappadocia.* Erga Edizioni.

Bilici, M. (2002). *Yeraltı Şehirleri Kazısı.* Paper presented at the 13. Müze Çalışmaları, Ankara.

Bixio, R. (2012). *Cappadocia: Records of the underground sites*. Archaeopress.

Bixio, R., & Castellani, V. (2002). Dispositivi di difesa nei sotterranei cappadoci. In R. Bixio, V. Castellani, & C. Succhiarelli (Eds.), *Cappadocia. Le città sotterranee* (pp. 265–278). Istituto Poligrafico e Zecca dello Stato.

Bixio, R., Castellani, V., & Succhiarelli, C. (2002a). *Cappadocia, le città sotterranee*. Istituto Poligrafico e Zecca dello Stato.

Bixio, R., Castellani, V., & Succhiarelli, C. (2002b). Derinkuyu, una città nel sottosuolo. In R. Bixio, V. Castellani, & C. Succhiarelli (Eds.), *Cappadocia. Le città sotterranee* (pp. 243–252). Istituto Poligrafico e Zecca dello Stato.

Bixio, R., & De Pascale, A. (2012). *Defensive devices in ancient underground shelters*. Paper presented at the International Symposium on East Anatolia-South Caucasus Cultures, Erzurum (Turkey).

Bixio, R., Yamaç, A., Galeazzi, C., & Parise, M. (2021). Artificial cavities of Turkey. Updating the map of anthropogenic cavities in the Mediterranean Basin. *Opera Ipogea 2/2021*, 27–46.

Bixio, A., Bixio, R., & De Pascale, A. (2023). *Göreme and Şahinefendi; history inside the rocks of Cappadocia*. Archaeopress Publishing Ltd.

Bobrovskyy, T., & Grek, I. (2015). *An artificial cave complex in Meskendir Valley near Goreme (Cappadokia) (in Russian)*. Paper presented at the VI international scientific conference 'Speleology and Spelestology', Nabereznye Chelny.

Bobrovskyy, T., & Grek, I. (2016). *The artificial cave complex in fairy chimneys in the vicinity of Göreme (in Russian)*. Paper presented at the VII international scientific conference 'Speleology and Spelestology', Nabereznye Chelny.

Bobrovskyy, T., & Grek, I. (2018). *A contribution to prayer places in central Anatolian cave shelters (in Russian)*. Paper presented at the IX international scientific conference 'Speleology and Spelestology', Nabereznye Chelny.

Bobrovskyy, T., Grek, I., & Klimishina, O. (2015). *A complex in a standalone fairy chimney in Zemi Valley, Cappadokia (in Russian)*. Paper presented at the VI international scientific conference 'Speleology and Spelestology', Nabereznye Chelny.

Brownson, C. (1922). *Xenophon, Anabasis: Books IV–VII*. Loeb.

Castellani, V. (2001). Acqua, acquedotti e qanat. *Opera Ipogea, 2*, 25–32.

Cooper, E., & Decker, M. (2012). *Life and society in Byzantine Cappadocia*. Palgrave Macmillan.

Demir, Ö. (1990). *La Cappadocia, culla della storia*. Ajans Türk Publishing.

Ergeç, R. (2003). Nekropolen und Gräber in der südlichen Kommagene. *Asia Minor Studien, Bonn Band 47*.

Erkmen, M. (1998). *Aydıntepe Yeraltı Şehri Çalışmaları*. Paper presented at the 9. Müze Kurtarma Kazıları Semineri, Ankara.

Faydalı, E. (1992). *Kavlaktepe Yeraltı Şehri Kurtarma Kazısı*. Paper presented at the 2. Müze Kurtarma Kazıları Semineri, Ankara.

Faydalı, E. (1993). *Kavlaktepe Yeraltı Şehri Kurtarma - Temizlik Kazısı*. Paper presented at the 3. Müze Kurtarma Kazıları Semineri, Ankara.

Galeazzi, C. (2013). The typological tree of artificial cavities: A contribution by the commission of the Italian speleological society. *Opera Ipogea, 1–2013*, 11–20.

Gilli, E., & Yamaç, A. (2015). *Water tunnels of Güvercinlik valley (Cappadocia - Turkey)*. Paper presented at the HYPOGEA.

Giovannini, L. (1971). *Arts of Cappadocia*.

Grek, I., Dolotov, Y., & Podruckaya, N. (2010). *An underground city in the Melendiz Valley*. Paper presented at the I international congress of speleology and speleostology, Naberznhye Chelny (Russia).

Grek, I., & Podruckaya, N. (2012). *Cave city in Shiraly lake crater*. Paper presented at the III international congress of speleology and speleostology, Naberznhye Chelny (Russia).

Gülyaz, M., & Yenipınar, H. (2007). *Underground cities of Cappadocia*. Nevşehir.

Haldon, F., & Kennedy, H. (2006). The Arab—Byzantine frontier in the eighth and ninth centuries: Military organisation and society in the borderlands. In H. Kennedy (Ed.), *The Byzantine and early Islamic near east*. Routledge.

Hauqal, I. (1964). *Configuration de la terre (Kitab surat al-ard)* (Vol. 1). Beyrouth-Paris.

Hawkins, J. (2024). *Corpus of hieroglyphic luwian inscriptions* (Vol. 3). Berlin.

Hild, F., & Restle, M. (1981). *Kappadokien (Kappadokia, Charsianon, Sebasteia und Lykandos)*. Verlag der Österreichischen Akademie der Wissenshaften.

Jerphanion, G. D. (1925–42). *Une nouvelle province de l'art byzantin, les églises rupestres de Cappadoce*. Paris.

Kempe, D. (1988). *Living underground*.

Kloner, A., & Zissu, B. (2007). *The necropolis of Jerusalem in the second temple period*. Peeters.

Mora, C., Balza, M., Bixio, R., & De Pascale, A. (2017). *A link between "Ancient Words" and the "Underground World": Cappadocian landscape, Rock-cut structures and textual evidence from Hittite documentation*. Paper presented at the HYPOGEA, Cappadocia (Turkey).

Mora, C., Balza, M., Bixio, R., & De Pascale, A. (2020). *Cappadocia, a "Hidden" Landscape: "Underground World" and neo-hittite evidence*. Paper presented at the 11th International Congress on the Archaeology of the Ancient Near East, Munich.

Okuyucu, D. (2007). *Derinkuyu Yeraltı Şehri (Derinkuyu Underground City)* (Ma). Atatürk Üniversitesi - Erzurum, Erzurum.

Ousterhout, R. (2017). *Visualizing community: Art, material culture, and settlement in Byzantine Cappadocia* (Vol. 46). Harvard University Press.

Özkorucuklu, H. (1991). *Aydıntepe Yeraltı Kenti*. Paper presented at the 2. Müze Kurtarma Kazıları Semineri, Ankara.

Öztürk, F. (2009). *Rock carving in Cappadocia from past to present*. Arkeoloji ve Sanat Yayınları.

Öztürk, F. (2017). Rock-cut Architecture. In P. Niewöhner (Ed.), *The archaeology of Byzantine Anatolia: From the end of late antiquity until the coming of the Turks* (pp. 148–159). Oxford University Press.

Ramsay, W. (1897). *Impressions of Turkey during twelve years' wanderings*.

Şahin, S. (1993). *Özlüce Yeraltı Şehri*. Paper presented at the 3. Müze Kurtarma Kazıları Semineri, Ankara.

Talbot, A., & Sullivan, D. (2005). *The history of leo the deacon: Byzantine military expansion in the tenth century* (Vol. XLI). Dumbarton Oaks.

Triolet, J., & Triolet, L. (1993). *Les villes souterraines de Cappadoce*. Torcy.

Triolet, J., & Triolet, L. (2002). L'organizzazione difensiva. In R. Bixio, V. Castellani, & C. Succhiarelli (Eds.), *Cappadocia. Le città sotterranee* (pp. 253–264). Istituto Poligraficoe Zecca dello Stato.

Urban, M. (1969). *Yeralti Schechri, Massenflucht Unter die Erde*. Verfasser.

Urban, M. (1973a). Das Rätsel der unterirdischen Städte Südostanatoliens. Dritter Teil: Maginotlinie der Frühgeschichte. *Vorland 8*, 205–212.

Urban, M. (1973b). Das Rätsel der unterirdischen Städte Südostanatoliens. Erster Teil: Der Befund. *Vorland, 6*, 150–153.

Urban, M. (1973c). Das Rätsel der unterirdischen Städte Südostanatoliens. Zweiter Teil: Geschichtlicher Rahmen und Deutungen. *Vorland, 7*, 174–181.

Wagner, J. (1982). Neue Denkmäler aus Doliche. Ergebnisse einer archäologischen Landesaufnahme im Ursprungsgebiet des Jupiter Dolichenus. *Bonner Jahrbücher, Bonn, 182*, 135–156.

Wessels, K. (2000). *Renovating Qanats in a changing world, a case study in Syria.* Paper presented at the International Symposium on Qanats, Yazd, Iran.

White, G. (1904). The cavate dwellings of Cappadocia. In H. Baum (Ed.), *Records of the past* (Vol. 3, pp. 67–73). Records of the Past Exploration Society.

Yamaç, A. (2017). *Underground settlements and cave dwellings of Agirnas village, Kayseri (Southeast Turkey).* Paper presented at the HYPOGEA, Cappadocia, Turkey.

Yamaç, A. (2019). Büyük Bürüngüz underground shelter (Kayseri – Turkey). *Opera Ipogea, 2–2019*, 65–76.

Yamaç, A. (2022). Some interesting underground cities and peculiar underground structures of Kayseri (Turkey). *Jerusalem Journal of Archaeology* (3), 70–107.

Yamaç, A., & Gilli, E. (2016). Underground shelters of Golgoli (Cappadocia - Turkey). *Opera Ipogea, 1*, 33–40.

Yamaç, A., & Okuducu, Z. (2017). *Underground hydraulic structures of gaziantep (Southeast Turkey): Livas and qastels.* Paper presented at the HYPOGEA, Cappadocia, Turkey.

Yamaç, A., & Tok, E. (2015). *Surveying some of the touristic underground cities of Cappadocia (Turkey).* Paper presented at the HYPOGEA Rome

Yamaç, A., & Tok, E. (2016). Doganli underground shelter (Capadocia - Turkey). *Opera Ipogea, 2–2016*, 57–62.

Yamaç, A., Tok, E., & Filikci, B. (2014). St. Mercurius underground city of Saratli (Aksaray-Turkey). *Opera Ipogea, 2–2014*, 37–46.

Yörükoğlu, Ö. (1989). *Underground cities in Cappadocia.* Ankara.

Yörükoğlu, Ö., Sevil, T., Taşcı, Z., & Türkmen, K. (1988). *Les Sites Souterrains de Cappadoce.* Ankara.

4 50 Significant Underground Shelters

Abstract

This chapter presents a detailed analysis of 50 underground shelters selected from the 253 structures listed in Sect. 3.6. The decision of which shelters to include on the list was not straightforward. On one hand, there were shelters that were clearly important but had not been studied and did not even have a plan. On the other hand, there were structures that had been studied extensively and even been the subject of articles but had no architectural or structural significance. In the face of such a contradiction, we were forced to make a mixed choice that included both groups. Therefore, unfortunately, the plans of some of the subterranean shelters described in this chapter are in sketch quality and do not have high-quality photographs. Although the majority of these 50 selected shelters are in Nevşehir and Kayseri provinces, we also tried to give examples from some other provinces that include ancient Cappadocia.

Keywords

Underground shelter · Underground city · Subterranean shelter · Artificial cavity · Cappadocia

4.1 Acıgöl Underground Shelter No 1

The underground shelter is located in the central part of Acıgöl District, within Nevşehir Province. The name *Acıgöl*, which translates to "Bitter Lake" in Turkish, is geographically and etymologically significant. Approximately 12 km to the south, at the juncture between the broad plain and the slopes of Erdaş Dağı, lies the village of Kurugöl, meaning "dry lake." The names of these locations, coupled with the regional topography, suggest the historical presence of a saline body of water similar to *Tuz Gölü* ("Salt Lake"), situated about 65 km to the west. This lake may be considered the ideal western boundary of Cappadocia. Tuz Gölü itself is a closed basin characterized by an average depth of 1 m and substantial evaporation, and it is also notable for its salt production.

At present, only specific portions of the Acıgöl Underground Shelter are accessible to visitors (Fig. 4.1). There are at least six distinct entrances to the subterranean complex, although only three remain visible. Certain sections of the settlement have either collapsed due to natural causes or become obstructed by soil displaced as a result of rainfall—ironically, a consequence of renovation efforts carried out several years ago. Visitors currently enter the site through a tall structure constructed above ground level. From this point, a tunnel, which was built approximately 50–100 years ago, descends beneath the road surface. At the end of this corridor lies the first chamber [2], which contains an operations room (M1) featuring a millstone door that once sealed the original entrance.

Adjacent to this chamber is a short, curved tunnel leading to the second room [3], with an aperture blocked by a dry-laid stone wall. A tunnel on the northwest side of the room connects to the exterior, though it is partially blocked. On the opposite side, a narrow, winding corridor—both low and narrow in form—contains niches presumably used for lamps and leads to a sequence of connected rooms. A brief corridor extending from the final chamber [4] is constructed from large stone slabs and is secured by another millstone door (M2). Light from outside penetrates the interior through cracks in the structure.

Due to the limited portion of the shelter currently accessible, the site's analysis remains inherently constrained. There is a clear need for further comprehensive investigation of the collapsed areas, in addition to a detailed study of the closure mechanisms and excavation directions (Bixio, 2012).

A. Yamaç and E. Tok, *Underground Shelters of Cappadocia, Turkey*, Advances in Karst Science,
https://doi.org/10.1007/978-3-032-16095-9_4

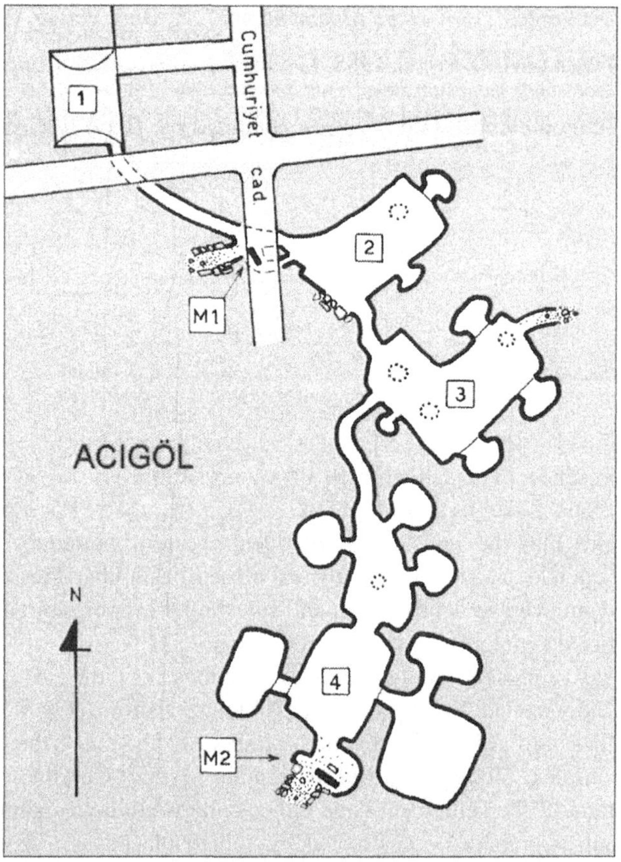

Fig. 4.1 Acıgöl Underground Shelter No 1 plan (Bixio, 2012)

4.2 Acıgöl Underground Shelter No 3 (Topada)

The underground shelter is situated 3 km northwest of Ağıllı Village and 9 km south of Acigöl. The Hittite inscription referred to as the "Topada Stele" is situated on a tiny hill elevated above the valley in this region. The underground shelter is accessed via the opening located approximately 100 m from the inscription. It is sometimes referred to as the Ağıllı Underground Shelter (Fig. 4.2).

The side adjacent to the stele contains several small rooms excavated along the front. The system TO 1 comprises a basic chamber [1] including lateral recesses, potentially serving as a stable. The exterior wall has fallen, along with the adjoining rooms that were connected by a short tunnel.

The TO 2 system comprises three interconnected rooms, each with separate external entrances. The room is currently accessible via an opening in the ceiling. We hypothesize an elevated ambient temperature, with the entrance obstructed. The configuration of the final chamber [3] appears to exhibit features typical of a church, including two lateral tombs.

The TO 3 system, situated on the opposite side, exhibits a more organized development, predominantly parallel to the outcrop's front. It possesses two entry points [1 and 2], which are partially obstructed by gravel deposits.

A brief corridor from room [1] connects to a third room, which is central to several interconnected rooms [4, 5, 6].

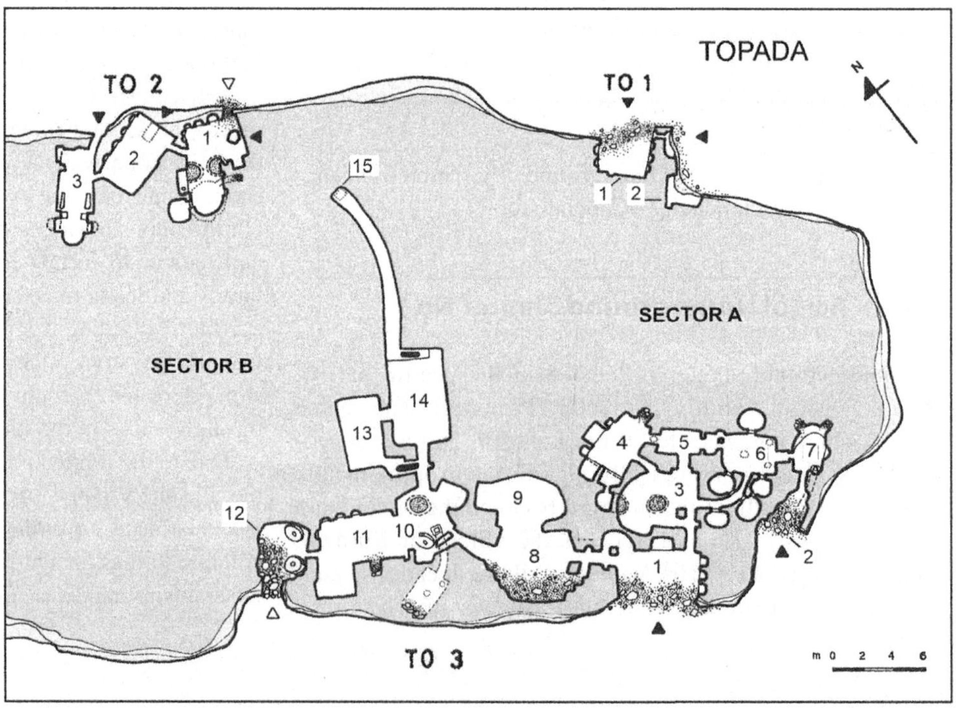

Fig. 4.2 Acıgöl Underground Shelter No 3 (Topada) plan (Bixio, 2012)

Circular-shaped warehouses and subterranean storage silos are present. The sixth chamber is linked to another room [7] that communicates with the exterior [2]. The latter features a lower duct (a) filled with earth, along with two other tunnels obstructed by landslides (b, c), similar to the tunnel (d) in room [4].

Upon reentering the initial chamber, a secondary brief lateral passageway extends to a succession of expansive rooms [8, 9, 10, 11, 12], aligned parallel to the hill's facade. The chamber [12] had an exterior entrance, now obstructed by rubble.

A perpendicular tunnel from room [10] extends to two chambers hewn into the interior of the hill [13, 14]. On the opposite side of the room, another passage extends in the same direction for approximately 12 m, terminating in an ascending pipe obstructed by a stone slab. It was likely in contact with the summit of the hill. A descending tube from the floor of room [10] leads to a small room [16] partially obstructed by collapsed rubble.

The interrupted sections imply additional advancements of the village within the rock outcrop. The duct in compartment [7] and the landslide in compartment indicate a more extensive bottom level; yet, the area of the settlement now examined is predominantly situated on a singular horizontal plane.

We see a significant concentration of shutting devices compared to the limited expansion of the underground network. The examination of the device locations indicates that the community comprises two distinct sections with different entrances. Sector A has access through room [2], while sector B's entrance is located in room [12].

The room [1] presently serves as a corridor connecting both sections; nonetheless, it is probable that this space once did not have external access. The opening would consequently occur from the collapse of the rock frame. This theory arises from the observation that the connecting tunnel between room [1] and room [8] consists of two operational rooms, positioned directly opposite each other, which would be rendered ineffectual if the initial compartment were readily accessible to adversaries approaching from outside. Millstone doors are absent, might thus protect one another and alternately the two sectors A and B. The devices are then duplicated to enhance the security of rooms [3] and [10].

Subsequent investigations will undoubtedly result in the identification of additional closure mechanisms designed to secure the entrance from outside area A, situated in room [2].

The presence of defenses at the entrance of room [12] in sector B is apparent due to the two millstone doors. The investigation of the defenses of the inner compartments [13] and [14] becomes pertinent. The millstone doors safeguarded the redoubt from threats in both directions (Bixio, 2012).

4.3 Ağırnas Underground Shelter

Located on the slope facing northeast on the left bank of the Koramaz Valley, at the southern entrance of the village of Ağırnas—Kayseri, this structure was excavated, as almost all in Cappadocia, in soft volcanic rock. It has five different entrances, three of which are blocked and all protected by millstone doors (Fig. 4.3). The first millstone door is located just behind what is today the touristic entrance. The room beyond this entrance, measuring 3.2 × 2.6 m, is connected to the inner church by a short passage on the opposite wall. It can be assumed that this first room had the function of narthex. The millstone doors in the tunnels right behind the church prove that the church was used in normal times apart from the raids.

The floor of the 6.60 m long naos of the Ağırnas Underground Shelter church contains several graves, the ceiling is barrel-vaulted, and the plan of the apsis is semicircular (Fig. 3.9). Church and narthex are in turn connected to the northwest with two other rooms, perhaps at the service of the clergy or for storage.

The large and long chamber on the east, measuring approximately 24 m in length and 4 m in width on average, is accessed via another short windy tunnel from the eastern side of the previous first chamber, perhaps protected by a millstone, now disappeared. This large chamber (or gallery), which develops with slight undulation on the north–south axis, constitutes the main body of the whole structure. To the west and east of this gallery, there are two additional chambers (halls).

The "table and sitting places" found in the side chamber to the east (east hall), excavated from the bedrock, represents an architectural element that can be found in numerous examples in Cappadocia. While a portion of the structure has collapsed and the lower floor has been exposed, the undamaged portion is still evident. In general, for Byzantine monasteries, the consistency of certain elements in their appearance may facilitate their identification as such. According to Svetlana Popovic, the presence of an enclosure wall, a church for communal worship, and a refectory for the taking of communal meals would suggest the presence of a monastic establishment (Popovic, 1998). Likewise, Spiro Kostof in his book *Caves of God* highlights two particular spaces within a carved complex, *the church* and *the refectory*, as being the main indicators of a monastic establishment in Cappadocia (Kostof, 1972). For the latter, this would mean the presence of a long rock-cut table, *trapeza* in Greek, and flanking benches.

These refectories are found in numerous rock-cut monasteries throughout the region, and several studies have been conducted on the topic (Lucas, 2003; Ousterhout, 2010, 2017; Öztürk, 2012). We point out that in some cases,

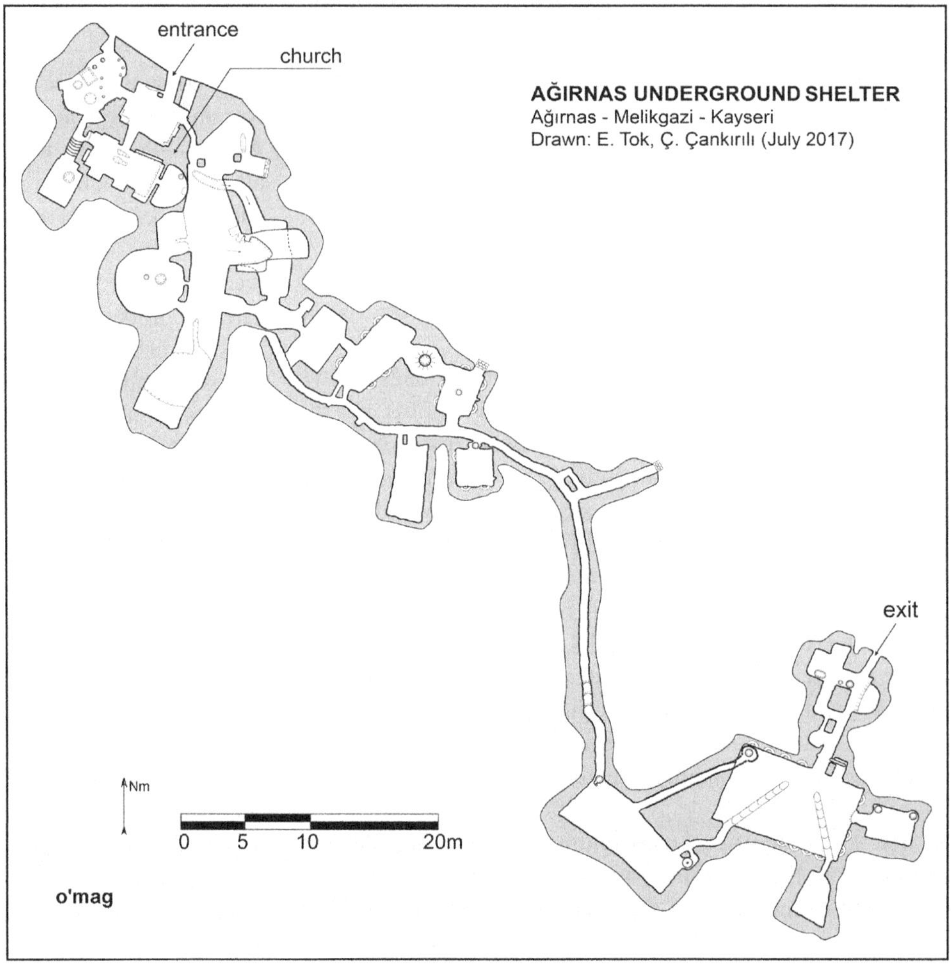

Fig. 4.3 Ağırnas Underground Shelter plan (drawn E. Tok, Ç. Çankırılı)

for example in the Göreme area, Ousterhout (2010, p. 97) thinks that several refectories were related to burial chapels and used for commemorative meals (*refrigeria*). Also, in the same section, opposite the hall where the refectory is located, there is another circular hall (west hall) with a 5.50 m diameter and a high ceiling. It has a huge chimney and a fireplace on the ground, suggesting the possibility that this facility was a kitchen. These kinds of spacious kitchens with high conical, pyramidal, or domical forms, called "chimney-vault" by Ousterhout, are also considered a distinctive element of the Cappadocian monastery or residence architecture (Ousterhout, 2005).

The other entrance on the northeast of the gallery has been recently closed. However, a millstone door defending the entrance still exists (Fig. 4.4). In the main tunnel extending for 60 m in the southeast direction from the refectory hall, there are 11 living areas/storage rooms.

Along the first segment the tunnel is sinuous, with rooms on both sides, and there are two points whose morphology makes one think of the housing for transverse millstone doors, now missing. These elements need to be better investigated.

Another door, almost in the middle of the tunnel and opening outside, has been blocked with stones. Though the millstone door, which should be 5 m inside a short tunnel from this entrance, does not exist today, the operations room is apparent, defending the free transit, in both directions, from possible external attacks.

From here the main tunnel extends to the south, connecting the two aforementioned sectors to a third cluster of rooms, accessible by a separate entrance/exit. The cluster consists of a central large rectangular chamber, 11 × 6 m, with two smaller side rooms, all carved with very neat workmanship. In front of this room there is an articulated space consisting of orthogonal corridors, probably reworked at a later time, making useless a hypothetical millstone door, placed after an elbow bend, that could defend the entrance. However, a second millstone is placed at the entrance of the chamber. Between the chamber and the end of the main tunnel there is another large room, even more internal (redoubt) level, connected by a stepped ramp and a short bent tunnel. From the middle of this room a 10 m long tunnel extends northeast, ending in a well to catch groundwater. The entrance of the redoubt from the central chamber

Fig. 4.4 A millstone door and operation room in Ağırnas Underground Shelter (photo R. Straub)

is defended by another millstone, now fallen. This multiple defense scheme and a well in the innermost room are often found in Cappadocian shelters.

As we have stated when explaining Mimar Sinan Underground Shelter, though the connection tunnels are blocked, the fact that the same underground defense structure can be reached under four different dwellings apart from the house of Mimar Sinan is proof that there is a very large underground shelter extending under almost the entire Ağırnas Village is certain.

The existence of another independent underground shelter excavated 200 m away from the village despite the existence of such a big defense structure, which extends under the entire Ağırnas Village and can be reached from all dwellings, can seem meaningless and unnecessary at first. The first explanation of this situation is that a thousand years ago, in this region, there could be another small settlement, which was a part of Ağırnas Village with no trace today, and the residents could have dug a different "defense structure" for themselves. But while all the centuries-old structures in this village are still intact, it is unlikely that even a trace of a different settlement 200 m away from the village remained and was completely destroyed. Another more interesting theory about Ağırnas Underground Shelter is that this relevant structure can be a religious structure. If we accept the two aforementioned structures as a refectory and kitchen, the existence of the church and of several chambers in this underground site brings forward the possibility that this complex was even a monastery in normal times (Yamaç, 2023, 2024).

4.4 Ali Dağı Underground Shelter

Even though it is named as "underground shelter" in the brochures, this structure located on the slope of Ali Dağı, 12 km from Kayseri center, is an architectural oddity by all means. A sketch plan of this structure was prepared by Yörükoğlu (1989). Commenting on this sketch in his book, Bixio writes:

> … the network of tunnels, as seen from the graphics, appears preposterous and, above all, unjustified. A topographic survey is expected to lead to shorten or change the plan. (Bixio, 2012)

Although the survey and mapping carried out by OBRUK Cave Research Group two years after these lines of Bixio reveal that the real image of the structure is quite different from the Yörükoğlu sketch, the "preposterous" status of the tunnels remains valid.

As a result of later survey, this structure, which was determined to have a total gallery length of 528 m, does not carry an "underground shelter" identity in any way when compared to all other known and surveyed underground shelters (Fig. 4.5). Even if we accept that the millstone door at the entrance of today's exit gallery was used to close the gallery and to defend the part that continues to the back of this gallery, behind this millstone, there is only a large chamber, a 180 m long tunnel, and a 60 m long cistern (Fig. 4.6).

It is unclear why it was necessary to dig a tunnel of this length to build a cistern. The southwest-facing branch of the circular tunnel in the middle of the long tunnel reaching the cistern was abandoned without much digging.

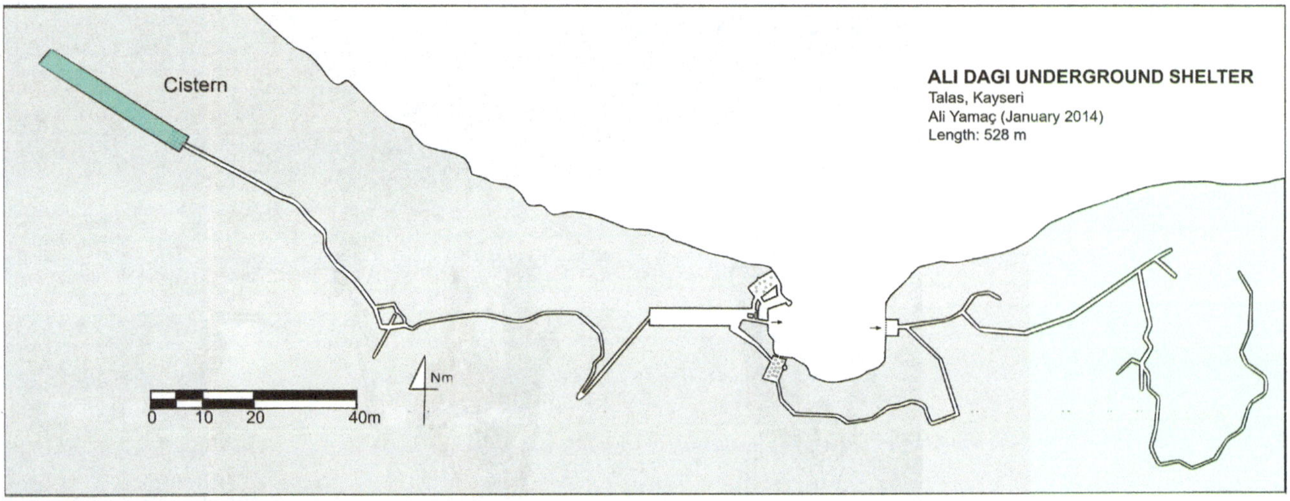

Fig. 4.5 Ali Dağı Underground Shelter plan (drawn A. Yamaç)

Fig. 4.6 Cistern inside the Ali Dağı Underground Shelter (photo A. Yamaç)

Again, although the first construction purposes of the three small rooms to the north of today's exit were a warehouse, kitchen, and church, it is certain that in the following centuries they were turned into dovecotes with windows opening to the outside. There is no room-like structure in the building that could be used as a residence in case of defense. On the other hand, there are no clogged galleries in the parts that continue to the east of the entrance which is closed today. In this section, all the tunnels seem to have been abandoned before they were completed as the potential threat has been dissipated (Yamaç, 2022).

4.5 Aşağı Hasinli Underground Shelter

The underground shelter is located in Ovakent, formerly known as Aşağı Hasinli, within Yozgat Province, approximately 15 km southwest of Boğazlıyan. The structure lies to the west of the present-day village.

The site encompasses two sections that are currently inaccessible due to flooding and sediment deposits, which hinder a complete examination of the complex (Fig. 4.7). The primary entrance to the first section, which extends for 84 m, is situated within a house on the surface [1], adjacent to a necropolis west of the village. Descending a series of steps, one reaches a spacious rock-hewn subterranean area comprising two adjoining chambers [2] and [3], both equipped with ventilation shafts. On the southern wall of the larger chamber [2], three tunnel openings are visible, indicating potential routes for further investigation. At the base of this wall, two millstone doors are located, though they are no longer in use and appear to have been displaced.

A structural collapse in another part of the complex provides access to the second section, which features a total length of 108 m and consists primarily of a long tunnel. This tunnel initially extends eastward in a gently curved alignment. Approximately 15 m from the beginning, it is protected by a millstone door that remains in its original operating room (M1). From this point, the tunnel turns

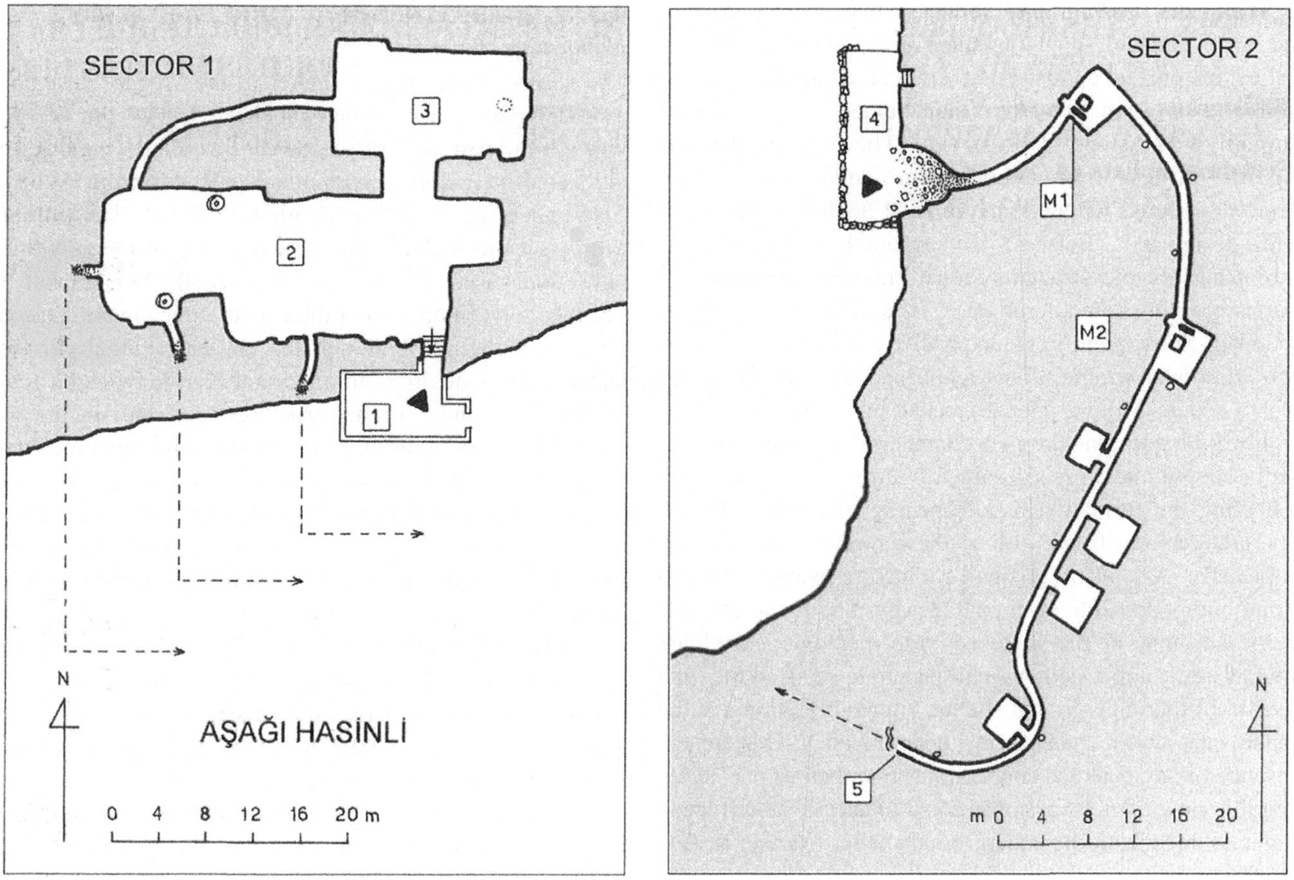

Fig. 4.7 Plans of sectors 1 and 2 of Aşağı Hasinli Underground Shelter (Bixio, 2012)

sharply and continues in a wide curve to the right, leading to a second operating chamber (M2), situated about 25 m from the first. This chamber also contains a millstone door, distinguished by a central opening.

Continuing southward, the tunnel extends for another 50 m. Modest rooms alternate along both sides of the passageway. However, flooding within the tunnel currently obstructs any further progress beyond this point.

The layout and features of both sections reflect typical characteristics of an "underground shelter." The two larger chambers [2] and [3], situated at the outermost section (first band), are interpreted as living quarters or areas intended for daily activities. The extended corridor [4–5], which is outfitted with defensive mechanisms and storage features, strongly resembles a connector tunnel that would typically lead to a redoubt—an inner, fortified space used as a temporary stronghold during times of attack.

It is reasonable to assume that additional subterranean chambers exist within the rock mass, particularly in the area between halls [2–3] and the tunnel [5]. These hidden rooms likely include essential features such as a water well. The three passages that extend from the main living chamber may have been designed to converge at a central point, facilitating quicker evacuation to the inner redoubt during emergencies. This zone was probably protected by the now-displaced millstone door, currently found within chamber [2].

The operational rooms may be located either within the interior compartments, at the tunnel entrances, or along the interconnecting corridors, such as in passage [5] of the second area (M1 and M2). This arrangement could have also supported the function of an escape route, allowing occupants to flee via an alternative path, potentially enabling them to confront or evade intruders.

The region between Aşağı Hasinli and Boğazlıyan contains a wealth of historical remains spanning multiple eras. These include thermal baths, Byzantine-era churches, Seljuk tombs, tumuli, and Hittite-period artifacts confirmed by hieroglyphic inscriptions (Bixio, 2012).

4.6 Belağası Underground Shelter No 1

Belağası Village is a small Armenian settlement located in the Gesi District of Kayseri. Currently, this entirely deserted village contains 28 rock-cut dwellings and two subterranean shelters.

The initial underground shelter, situated 10 m east of the water channel supplying the village from the Ötedere Valley, features an entrance that has been enlarged by a substantial excavation from a treasure hunter and is accessed through a millstone door in situ. The tunnel, measuring 70–80 cm in width and 90 cm in height, extends 42 m from this point without alteration (Fig. 4.8). The structure, designated "Belağası Underground Shelter No 1," exhibits a notable structural transformation following the extensive tunnel and terminates in a relatively spacious chamber. Subsequent to the expansive structure, there exist two additional sizable rooms featuring windows that overlook Ötedere Valley, interconnected by a staircase. It is highly probable that the two chambers are remnants of a wall collapse and were subsequently utilized as settlements following the formation of an opening to Ötedere Valley. The architectural framework of these rooms is remarkably impressive: A total of 24 small rooms are evenly arranged along both sides of a 34-m passageway. The tunnel located at the terminus of the structure, with a 60° slope inclined downward, is congested. A small peephole exists at the final section of the underground shelter, situated 32 m above the cistern and water channels in Ötedere Valley. Despite the absence of any detected connections from below, it is likely that this final tunnel was constructed to access subterranean water sources from the underground shelter (Yamaç & Tok, 2015b).

4.7 Belağası Underground Shelter No 2

The second underground shelter in the village, designated as "Belağası Underground Shelter No 2," is accessed through the rear wall of a rock-cut dwelling. This dwelling is located on a slope, approximately 80 m southwest of the first underground shelter mentioned above. The entrance to this second underground structure is positioned within a house measuring 12 by 9 m. However, the house itself has been largely filled with rubble over time, making access more challenging. Although the entrance tunnel currently lacks a millstone door, an important feature typically found in defensive underground structures, the shaft in the wall where the door would have once been fitted remains intact. This indicates that the shelter originally had a more secure entrance, likely designed to be sealed off when necessary.

As one moves further into the shelter, a distinct transformation in architectural style becomes evident. The initial design observed at the entrance gives way to a unique and somewhat unconventional layout. The subterranean structure follows a markedly different excavation approach compared to the first underground shelter. In this case, rooms have been carved on both the right and left sides of the main tunnel, in a manner that almost appears upside-down when compared to more traditional underground settlements. This unusual arrangement contributes to the shelter's distinctive architectural character (Figs. 4.9 and 4.10).

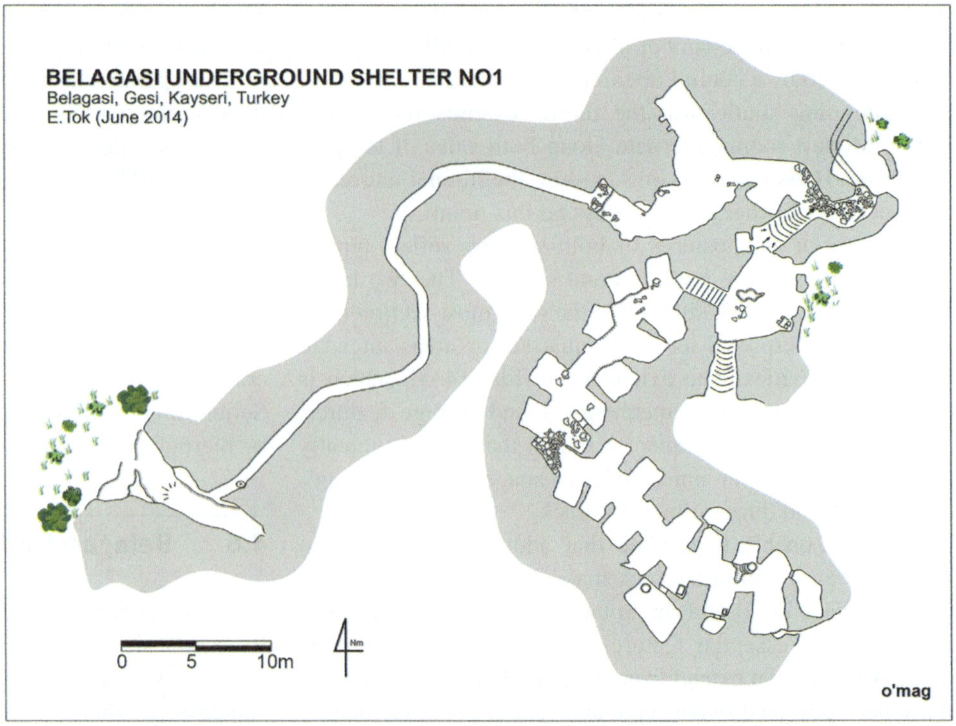

Fig. 4.8 Belağası Underground Shelter No 1 plan (drawn E. Tok)

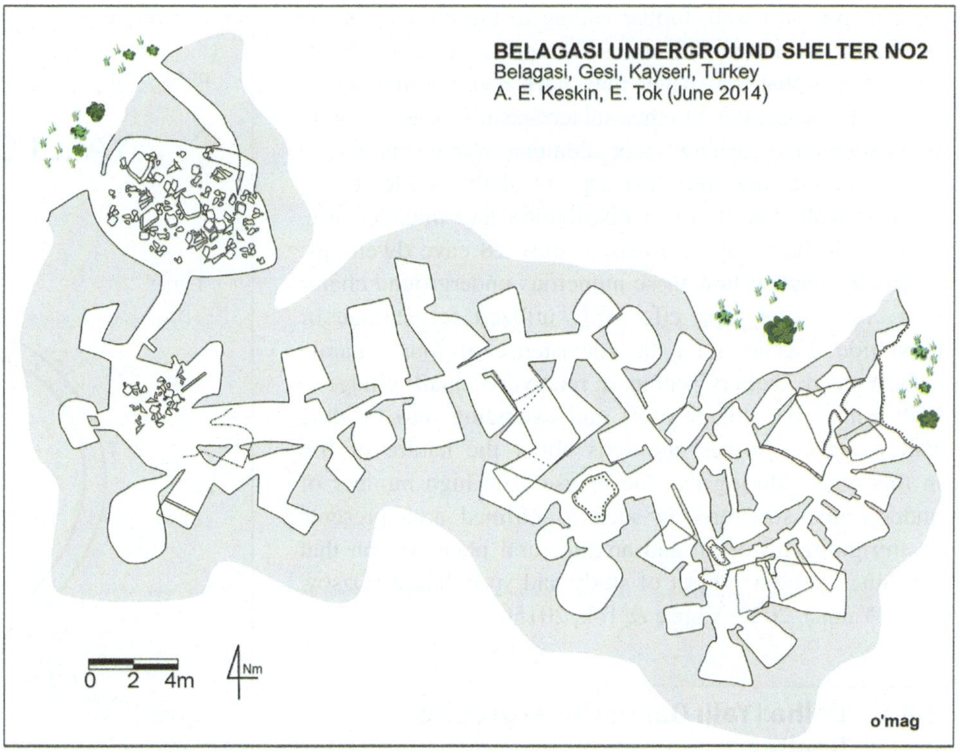

Fig. 4.9 Belağası Underground Shelter No 2 plan (drawn A.E. Keskin, E. Tok)

Fig. 4.10 Belağası Underground Shelter No 2 (photo A.E. Keskin)

The total number of rooms within this underground shelter is remarkably high, reaching an impressive 52 chambers. However, the majority of these rooms are quite small, generally measuring only 2–3 m^2. This creates a maze-like internal configuration, which is an intriguing and uncommon architectural phenomenon. Despite its relatively moderate overall dimensions, this underground shelter stands out for having one of the largest numbers of rooms recorded in Anatolia. It is comparable in complexity to some of the most well-known underground cities in the region, such as the vast subterranean shelters of Derinkuyu and Kaymaklı.

Similar to Belağası Underground Shelter No 1, this second shelter also features an opening that leads to Ötedere Valley. This opening, however, has been partially obstructed due to the collapse of a wall, further adding to the mystery of the site's past use and modifications. The presence of this opening suggests that the shelter may have had external access routes or connections to other subterranean spaces, possibly for ventilation, escape routes, or additional storage purposes.

One of the most puzzling aspects of this settlement is the practicality of its room distribution and intended use. Given that the village consists of only 28 cave dwellings, it remains unclear how these numerous underground chambers could have been effectively utilized for storage or habitation. Furthermore, the coexistence of four separate subterranean shelters within a relatively small village—both within the village itself and extending into Ötedere Valley—raises further questions about the nature of life in this region during its time of use. The high number of underground structures in such a confined area presents an intriguing historical and architectural phenomenon that continues to be a subject of study and speculation (Özsoy, 2017; Yamaç, 2022; Yamaç & Tok, 2015b).

4.8 Belha (Yallı Damı) Underground Shelter

Located near the village of Özkonak in the Avanos district, the rocky monastery of Belha comprises two subterranean complexes: Ayazma and Yallı Damı. The latter is positioned approximately 500 m away, on the western flank of the valley. The entrance consists of two low openings, each around 1.30 m in height, which lead into a large chamber with a ceiling height of 1.90 m and two adjoining rooms, labeled 1 and 2. From this point, one may proceed via two separate routes that comprise twin corridors cutting through the heart of the rocky hill. The first route is interrupted by an operating room (3), fitted with a millstone door (M1). Proceeding further, one arrives at room 4, which is secured by an additional defensive mechanism (M2). The corridor just traversed ranges in height from 1.10 to 1.35 m, while the ceiling height in the chamber measures 1.80 m. Room 4 grants access to room 5 via two pathways, one of which, located in the northwest, is currently obstructed by accumulated earth. From room 5, it is possible to exit through the second tunnel, which follows a winding route beginning at a height of 0.75 m and rising to 1.30 m. Notably, this passage appears to lack any defensive closures.

The architectural plan clearly indicates the specific locations of lamp niches carved along the two tunnels. However, the presence of a dual-passage system is difficult to interpret, as it does not exhibit the conventional characteristics of an "escape route," which underground shelters typically incorporate. In this case, the route seems inefficient, given that it circles back close to the original point of possible attack.

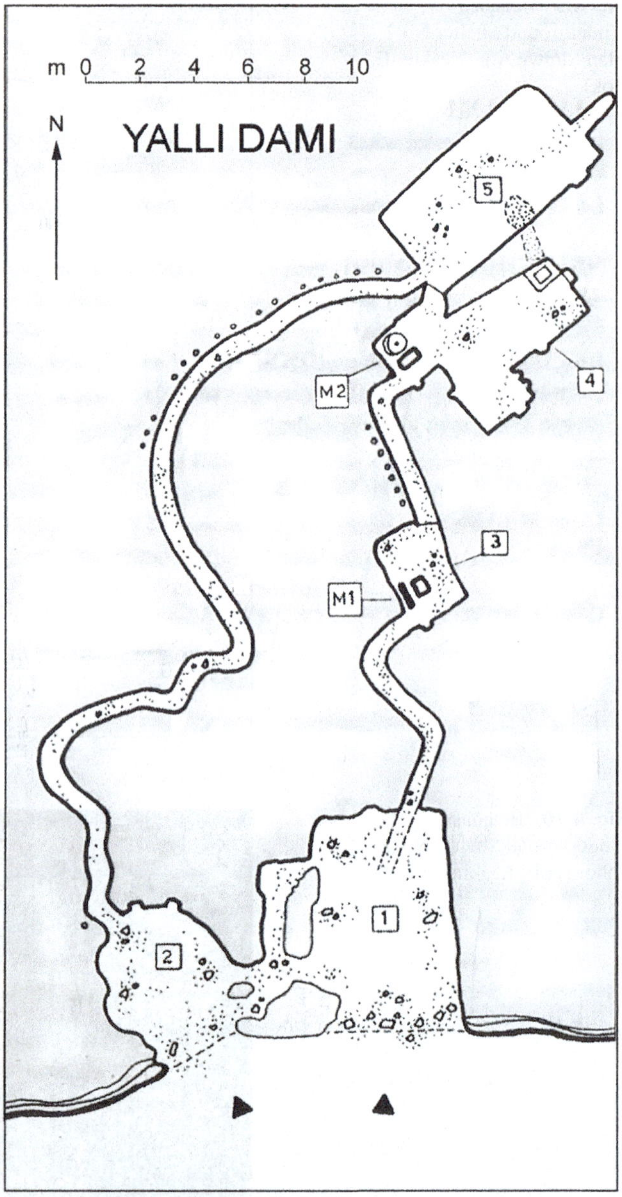

Fig. 4.11 Belha (Yallı Damı) Underground Shelter (Bixio, 2012)

This interpretation is further supported by the direction in which the tunnels were excavated, which suggests that the construction process did not begin with a tunnel from the exterior to rooms 4 and 5, followed by a separate tunnel leading from the interior outward. Instead, it is more likely that both tunnels were excavated simultaneously by inhabitants who, upon detecting the transmission of sound, abruptly redirected their digging efforts. The two tunnels then gradually converged, resulting in the inner chambers—rooms 4 and 5—becoming tangent to each other despite a vertical difference exceeding two meters (Fig. 4.11).

The reasoning behind the substantial fortification of one corridor—characterized by two millstone doors and sharp changes in direction that increase the effectiveness of the defensive structures—remains unclear. In contrast, the other corridor is minimally protected, its defense seemingly reliant only on its limited size, which would force invaders to crawl, thereby reducing their mobility and increasing their vulnerability to attack. However, since rooms 4 and 5 are internally connected, the weaker protection offered by the second corridor would render the defensive measures of the more secure corridor less effective. It is possible that the less fortified tunnel was simply left incomplete, and that the installation of two additional operating rooms was planned at the two sharp directional changes, which otherwise remain difficult to explain. This would suggest that the construction was never completed, either due to an attack that forced the abandonment of the project or because the threat in the area had subsided.

It is also noteworthy that the design of Yallı Damı suggests it was not intended to serve as a refuge for the nearby Belha rocky monastery, as that complex already possesses its own subterranean shelter.

4.9 Büyük Bürüngüz Underground Shelter

Büyük Bürüngüz Village, located 22 km east of Kayseri, at the starting point of the Koramaz Valley, has always been the largest settlement in the valley. According to the Ottoman records of 1500, there were 99 houses in the village (Inbaşı, 1993). Unlike the other six villages located in the Koramaz Valley, there are no rock-cut structures or cliff settlements around Büyük Bürüngüz.

During the project of OBRUK Cave Research Group, we have been informed about the existence of an underground shelter in this village accessible from the gardens of the houses. We note that there are defensive structures similar to this in various settlements of Cappadocia. These singular structures, dug beneath the houses, have been connected as a result of the growth of the village and formed a defense network extending beneath the entire inhabited area. Derinkuyu and Kaymaklı underground shelters, located 70 km southeast of Kayseri, have been formed as a result of the connection of individual defense systems developed on several underneath stories (Bixio, 2012). Likewise, beneath Ağırnas Village, located six km northwest of Büyük Bürüngüz Village, there is an enormous underground shelter extending beneath the entire village (Yamaç & Tok, 2015a).

First underground shelter of Bürüngüz, the entrance of which started with a stone masonry tunnel, continues with eight consecutive storages. The main underground shelter begins after passing the eighth storage room. Interestingly, the millstone door of the first room in the underground shelter defends toward the continuation of the tunnel. Presumably, the rooms that currently appear as warehouses at the entrance were originally rooms of the original defense structure and most probably had been converted into warehouses. The second room had a two-way defense system both toward the entrance to the first room and the continuation of the tunnel and had a double millstone door (Fig. 4.12).

The defense toward the inner tunnels proves that there are more connections of the tunnels with other houses. This second room that is mentioned should be a redoubt, i.e., a "last defense room" that defends against possible

Fig. 4.12 Separate tunnels in Büyük Bürüngüz Underground Shelter just after the first entrance (photo A.E. Keskin)

enemies coming from both directions. The redoubts, which are frequently encountered in the underground cities of Cappadocia, are almost always in the last room. On the other hand, this is the first time we saw this formation at the beginning of a tunnel in this underground shelter. The last part of the underground shelter looks like a labyrinth due to the net-like tunnels, most of which have been closed by debris. The fact that at least three millstone doors were defended toward both the entrance and toward the continuation of the tunnel, even in the last section, shows us that the underground shelter continued after the last point.

After the first search that was done in Büyük Bürüngüz Village, we found another entrance. The entrance of this second underground shelter was close to the first structure we studied and similarities begin with the underground storage rooms. The tunnel of this second underground shelter which continues for 41 m, starts from the south of the last warehouse. The first millstone door that we encountered in the tunnel defends toward the entrance. Two more millstone doors protect the entrances in the four-room system accessed from the main tunnel. The other tunnel, which continues to the west from the same room, is 54 m. The millstone door that is located in the last room of the tunnel, which would be protecting the room toward the western tunnel, is the largest one we have ever encountered in the entire Kayseri region with a diameter of 1.85 m and a thickness of 40 cm.

Thus, we explored and mapped two different underground cities under Büyük Bürüngüz Village. The first underground shelter was 701 m, the second underground shelter was 572 m. Apart from the entrances being used in our research, we found two more entrance or exit points of the opening to the surface. Therefore, we have four different GPS coordinates to enter these underground shelters. When we put these coordinates into Google Earth, we noticed that these two underground shelters intersect at one point. Afterward, we examined these intersection points and find out that both were filled with rubble. We had cleaned the rubble in both tunnels and connected these two underground cities. To conclude, the total studied length of this underground defense system reached 1.273 m (Fig. 4.13). The fact that it is the longest underground shelter ever explored in the entire province of Kayseri, and that a total of 27 millstone doors found in this defense structure is a record for this region. On the other hand, we know that this underground network that has been measured and mapped to date covers only a small part of its original part. Twenty-three different blocked tunnels prove that this structure most probably extends under the entire village (Yamaç, 2019).

4.10 Çıralı Sinkhole Underground Shelter

Çıralı, located on the border of Aksaray–Konya provinces, near Karapınar, is home to the most interesting underground shelter in all of Cappadocia. Here, the walls of a sinkhole were excavated to create a cliff settlement and a shelter (Fig. 4.14). This semi-desert region, known as the Obruk Plateau (Obruk—Sinkhole in Turkish), has dozens of sinkholes, some of which are hundreds of meters in diameter.

Scientific studies show that the resistance of Neogene limestone formation has decreased due to the continuous withdrawal of groundwater for agricultural production in the region. This has led to the formation of new sinkholes in the region as a result of collapses over the past decades. It is known that groundwater fell by 24 m between 1983 and 2008.

The Çıralı Sinkhole features entirely vertical walls and possesses a singular descent from the north. A well-preserved pavement on the upper section of the path indicates that the settlement was adequately developed at some point in its history. The cliff settlement comprises various independent caves, with only occasional instances of small complexes containing multiple structures. The most distinctive cavity within the complex is located in the southern wall of the sinkhole. This system features three exits that lead to the surface. The diverging, expansive tunnel penetrates 65 m into the rock mass before making a right-angle turn to the east and proceeding an additional 90 m in that direction. Two substantial excavations obstruct the progression of the tunnel in this segment. The majority of these excavations are conducted by treasure hunters. The tunnel extends eastward for an additional 22 m from this point. The basalt millstone door, measuring 1.4 m in diameter, is not composed of the rock found in this shelter or even within this region. Due to the destruction of the millstone door and the absence of grooves in the walls of these doors, which are not retained in this type of niritic limestone, the direction in which the door provided defense remains undetermined.

The tunnel ascends merely a few meters from its entrance, and this elevation change is inadequate to attain the surface level. It was presumed that this incline would direct the tunnel to the subsequent depression, which is shallower and less extensive than the sinkhole. Satellite imagery indicates that the tunnel ascends along the general gradient of the valley adjacent to the sinkhole, where a small settlement currently exists.

Flanking the tunnel are small niches and depots, likely intended for material storage. Additionally, there are three minor branches located in the vicinity behind the millstone

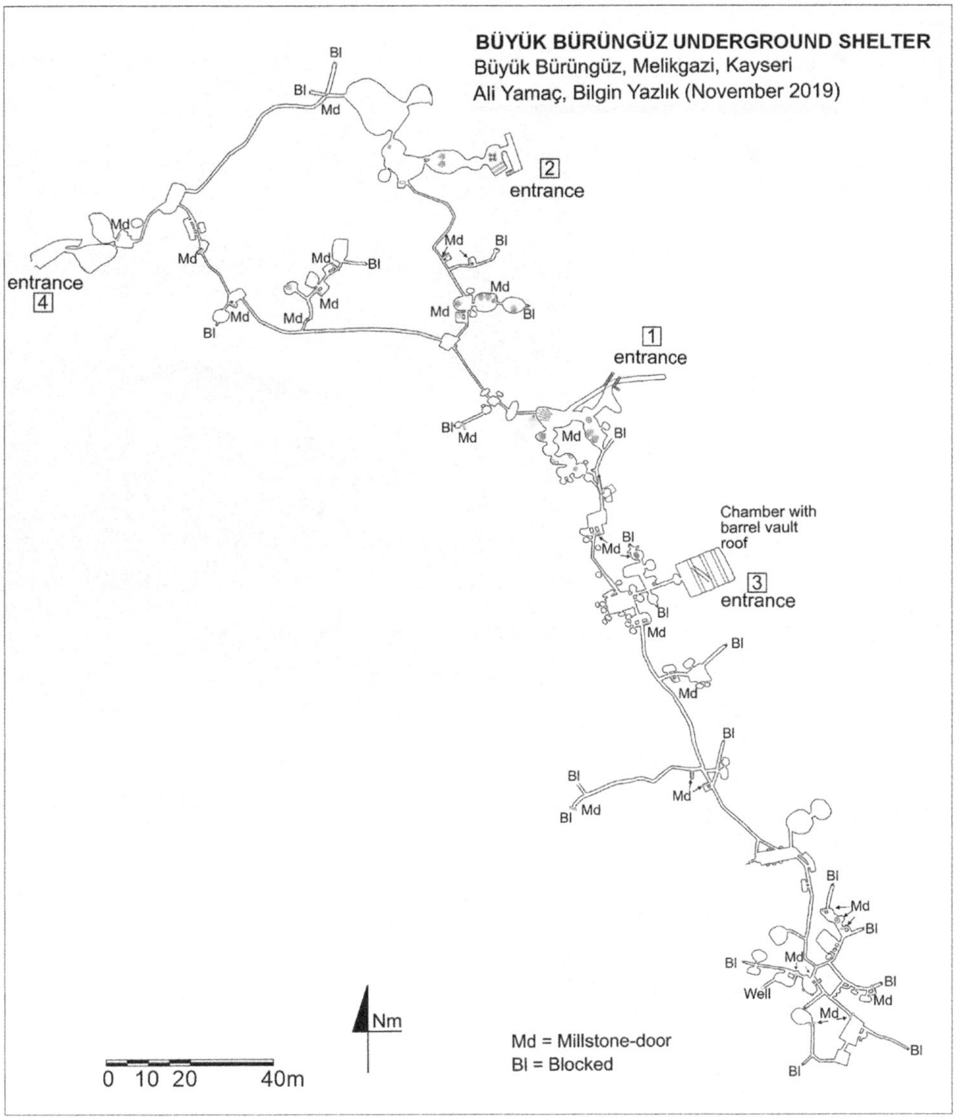

Fig. 4.13 Büyük Bürüngüz Underground Shelter plan (drawn A. Yamaç, B. Yazlık)

door. The branch nearest to the exits constricts and ascends in a southwesterly trajectory. This passage leads to a shaft, with two small rooms located on the upper floor. A separate vertical shaft terminates at the obstructed lower level.

The Çıralı Sinkhole Shelter encompasses an unresolved paradox. Under typical circumstances, this shelter is anticipated to serve as a defensive edifice for the inhabitants of the cliff settlement within the sinkhole. Nonetheless, the potential that this tunnel serves as the sole access point to the settlements within the sinkhole must not be overlooked. Individuals residing in an external settlement may have sought refuge in this settlement via the tunnel during an assault (Fig. 4.15) (Dolotov, 2014; Grek & Podruckaya, 2012).

4.11 Çomaklı Underground Shelter

This subterranean structure, situated in Çomaklı Village of Kayseri/Develi District, represents one of the most expansive of the 43 underground cities that have been surveyed during the "Kayseri Underground Structures Inventory Project." Its total length is 918 m.

It can be assumed that the underground shelter consists of two distinct sections. The northern section, extending from the entrance, consists of a series of interconnected rooms, while the southern section is traversed by a single tunnel that branches into two chambers.

The tunnel, which extends for a further 76 m in a southward direction, has an average width of 60 cm and a height

Fig. 4.14 Cliff settlements in the Çıralı Sinkhole, Konya Province (photo A.E. Keskin)

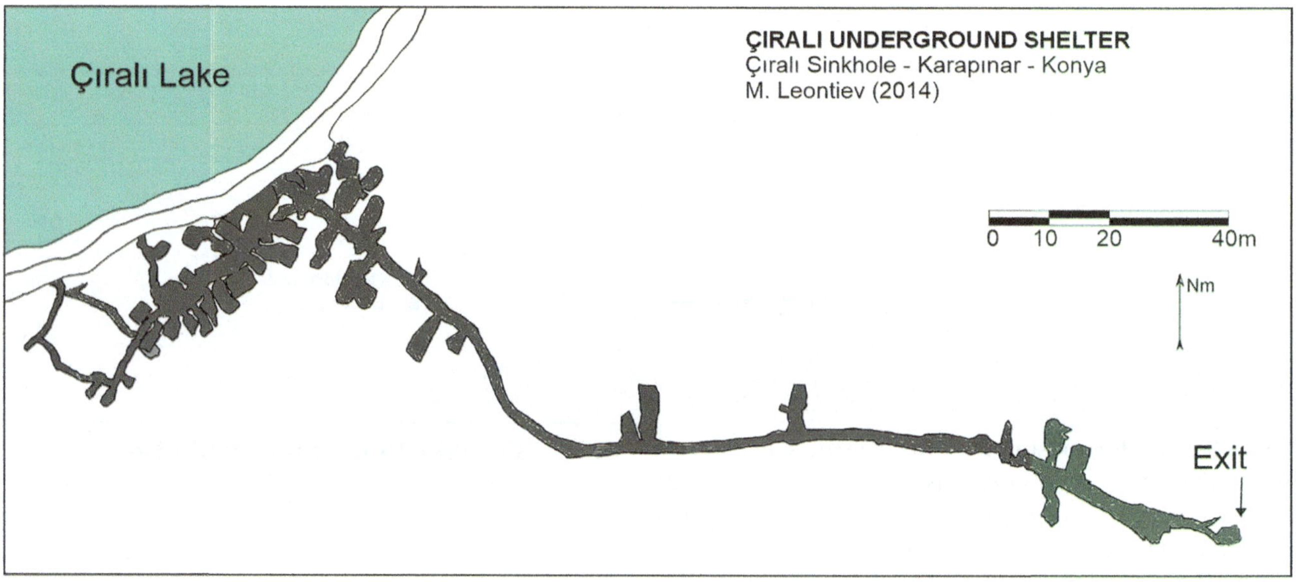

Fig. 4.15 Çıralı Underground Shelter (drawn M. Leontiev)

of 70–80 cm. It contains three small rooms in its final section. Subsequently, the continuation of the tunnel bifurcates into two. Despite the presence of a millstone door to defend the initial branch, the passage is obstructed by debris after a short distance. Similarly, the second branch is obstructed 7 m after the fork.

In contrast, the rooms in the northern part are mostly connected to each other by short and relatively wider tunnels. There are nine different rooms and an upper and a lower floor in this section. Most of these rooms, some of which are larger than 50 m^2, have silos on their floors.

A total of 12 of the tunnels in the northern section, which span three floors, are largely obstructed by debris. A total of seven millstone doors were identified within the sections of the shelter.

It was proposed that the underground shelter could be described as "abandoned without being finished" due to the presence of unexcavated tunnels at the entrance and in

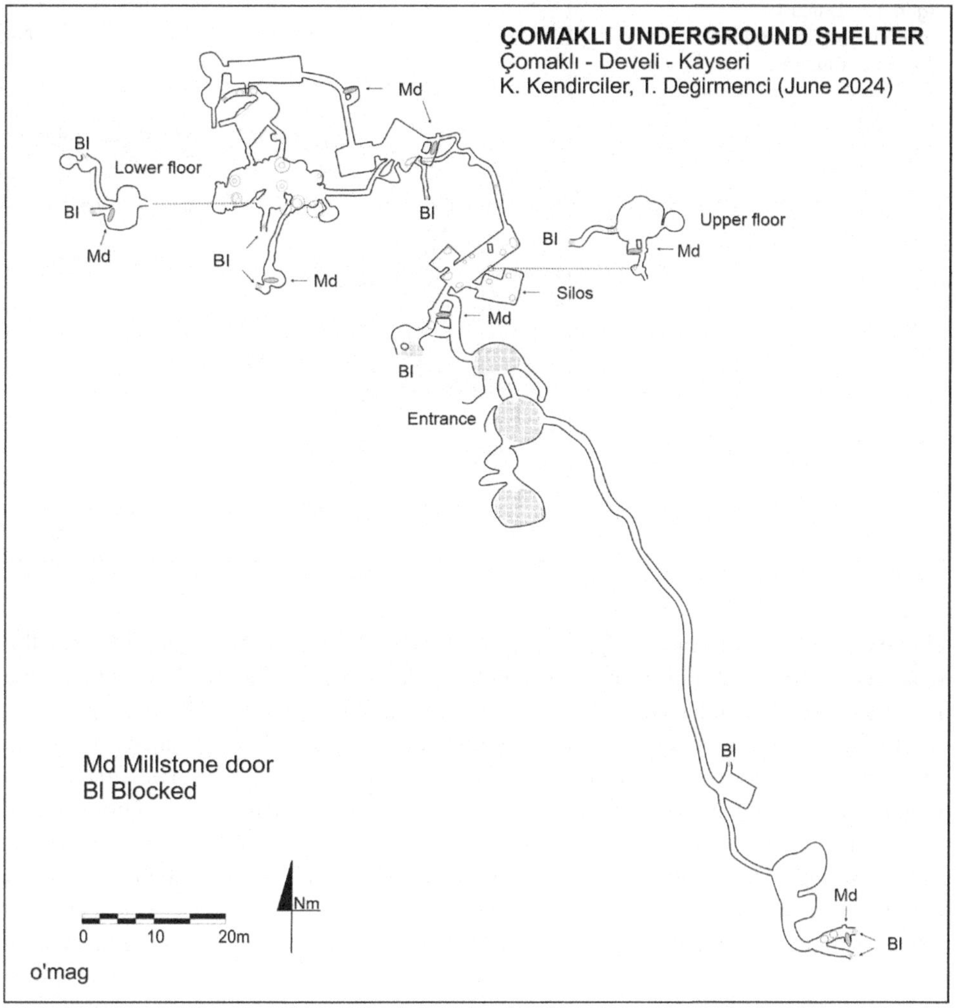

Fig. 4.16 Çomaklı Underground Shelter plan (drawn K. Kendirciler, T. Değirmenci)

the north. However, the heavy soot in all the lamp niches carved into the walls indicates that the structure was in use (Fig. 4.16).

4.12 Danalık Underground Shelter

The Danalık Underground Shelter is situated in Akdere, to the south of Çömlekçi Village in the Talas district. It is accessed via a narrow opening in the valley slope. Upon entering the tunnel, the team immediately observed the operation room and a groove on the opposite wall. However, there was no evidence of a millstone door. The tunnel culminates in a chamber that is equipped with subterranean silos at each corner.

The narrow tunnel, located to the west of the aforementioned chamber and extending in the same direction as the entrance, becomes blocked after a distance of 14 m. The fact that it extends in the same direction as the main entrance renders it implausible that this is an emergency escape tunnel. It is possible that the millstone door, which should be located within this tunnel, may be buried beneath the accumulated rubble.

This initial chamber gives access to three substantial rooms, which are linked to one another in the northwest, as well as another tunnel safeguarded by a millstone door. This then leads to a vast hall measuring 30 × 8 m. Two significant rooms with deep niches in their walls are accessed by a short and elevated tunnel on the northwest wall of the hall.

The other chambers on the southeast wall of the aforementioned large hall are of particular interest. In this section, there are two rooms and two small chapels. The continuation of the aforementioned large hall leads to another room after a lengthy tunnel, the entrance to which is protected by a further millstone door. At the eastern extremity of the aforementioned chamber, another room is accessible via a brief ascension to a short shaft. It can be reasonably deduced that this room, which is only accessible via the aforementioned shaft, is a redoubt.

A short tunnel continues downward in a slope to the south of the lower room and connects to another large chamber filled with leaking underground water. The

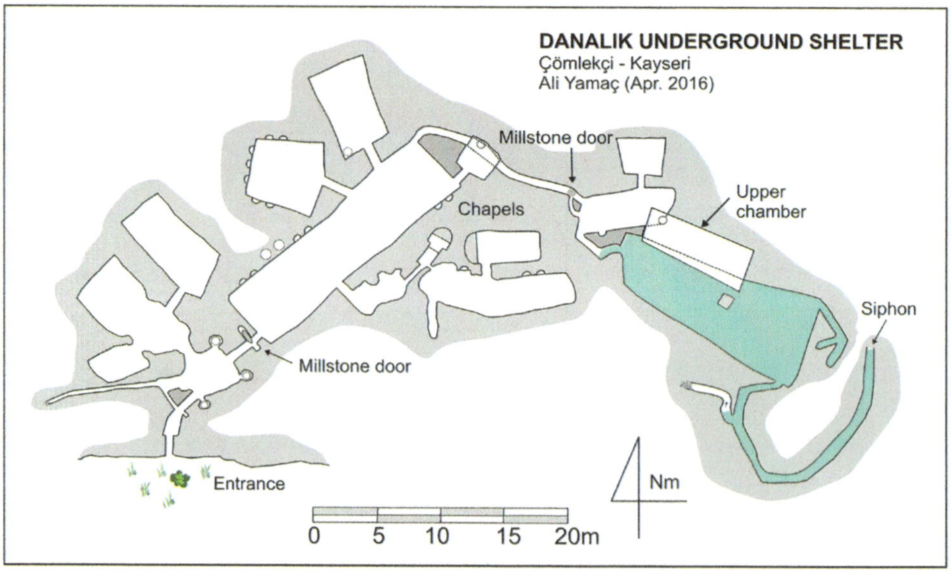

Fig. 4.17 Danalık Underground Shelter plan (drawn A. Yamaç)

chamber, with dimensions of approximately 17 × 6 m, was observed to be filled with water to a height of 80 cm during the research conducted in April. The continuation of the tunnel, which continues in a semicircular trajectory to the east of this room, could not be followed as it was completely filled with water (Fig. 4.17).

The fact that Danalık Underground Shelter is located on a deserted valley wall, two kilometers away from both Çömlekçi in the north and Alakuşak villages in the south, where there are no remains, remains unexplained.

4.13 Derinkuyu Underground Shelter

The many entrances of this underground complex are scattered among the buildings of Derinkuyu Village, 29 km south of Nevşehir. It was accidentally discovered in 1963 and partially opened to the public in 1965 by the General Directorate of Antiquities and Museums. The same year it was mentioned for the first time by Mahmut Akok and Hikmet Gürçay (Akok & Gürçay, 1965).

The underground complex of Derinkuyu is an extended structure consisting of multiple systems, some of which are partially explored and, however, are very poorly documented. Most of the underground network is still under investigation and in any case only marginally detected.

Derinkuyu consists of numerous rooms carved into tuff. The area of 2.500 m^2 is open for visitors. It covers an area of 4.000 m^2—including the uncleaned areas. It is not clear that how many floors the underground shelter has. Only 7 floors of the underground shelter were cleaned and it is believed that it might have 12 floors. The depth of the underground shelter is not known. Considering the ventilation pipe near the entrance it is 40 m deep. However, the well located at the seventh floor indicates that it is about 85 m deep. Today, it is entered through the corridor-like entrance which is used as the exit of the underground shelter. The first floor of the underground shelter consists of stable, winery, missionary school, and baptistery. The stable, which is one of the largest rooms of the underground shelter, comprises two large rooms built one within the other. There are small holes for water and feed in these rooms. The winery consists of two sections. The rectangular planned room called "missionary school" is accessible through the corridor in the stable. The baptistery is located over the entrance of this room and it is carved into the present floor. The corridor originating from the east of the stable at the first floor links the first floor to the second floor and the second floor to the fourth floor. The large area at the center of the second floor functioned as kitchen. The floor furnace is located at the center of the kitchen and there are niches used as shelves for supplies on the walls. Next to the kitchen, the winery consisting of two sections locates and there are silos on the east of the kitchen. The third floor of Derinkuyu underground shelter consists of the ventilation pipe which reaches all the floors. The tunnel originating from the first floor continues to the second floor and then turns the south and reaches the fourth floor located on the exit route of the underground shelter. At this floor, there are sitting-sleeping rooms and silos. The fifth floor consists of the platform where the tunnel comes from the third floor ends, the ventilation pipe, the rooms linked to each other and the beginning of tunnel linking the fifth floor to the seventh floor. Along the tunnel, there are five units. Two of them are door rooms and three of them are the security points to control the tunnel. The seventh floor is the largest floor of the underground shelter. This floor comprises of gathering hall, burial chamber and church. After the church, a hall and a well are located. The church in cross plan is located on the southwest of the gathering hall. The eighth floor consists of a small room with a ventilation pipe inside (Gülyaz & Yenipınar, 2007; Okuyucu, 2007).

Currently, the accessible entrances lead to multiple levels of underground structures and tunnels, fully excavated below the surface. Based on depth measurements and studies, there are three main levels: the first from 0 to −10 m (epidermal), the second from −10 to −20 m (intermediate), and the third reaching depths of −45/−55 m (deep).

Several systems, investigated in 1995, have not been previously reported.

D6) Ismet Aktepe Evi: Located near the Byzantine church (Büyük Kilise), this vast underground complex extends beneath the neighborhood, centered around a deep, debris-filled shaft estimated at seventy meters.

D7) Mümin Hoca Evi: An uninhabited building with a blocked entrance. Partial exploration revealed interconnected rooms and a millstone door in the closed position. A passage continues but remains unexplored.

D8) Hasan Demirel Evi: Adjacent to the previous system, with closed tunnels.

D9) Mustafa Sen Evi: A small dry-stone building grants access to horizontal tunnels beneath surrounding homes. A long, steep tunnel leads to a depth estimated at seventy meters, ending in a debris-blocked passage with signs of significant defensive structures. The entrance was later untraceable, possibly collapsed or intentionally sealed.

According to local sources, these three networks connect to a central shaft, forming an extensive underground settlement.

Unidentified Zone: 5 km west of Derinkuyu, in the Suvermez direction, an underground structure exists. The entrance leads into a large rock-carved room with multiple blocked passages.

Tourist Area: Derinkuyu 1

According to Demir (1990), the explored tourist area covers 2500 m^2 across eight floors, reaching −55 m. However, up to 18–20 additional levels remain unexplored, spanning 4 km^2, potentially sheltering 10,000 people. Derinkuyu features 52 large ventilation shafts, some reaching 85 m. Two main shafts exist in the tourist area: one remains unexplored, while the other extends 48 m but is filled with debris. The upper floors likely housed daily activities (kitchens, mills, wineries, and latrines), while the lower levels served as refuges, secured by millstone doors. A cross-shaped room at the lowest level is sometimes mistaken for a church.

Several corridors extend from the tourist area, leading to blocked entrances and other sections. Observations indicate a spiral development around a central vertical shaft, with tunnels intersecting at different levels. There are four levels across three bands:

Belt 1 (epidermal): Closest to the surface, containing the tourist entrance and multiple now-blocked access points. It features a large 20-m-long room with benches, possibly a church or refectory. Another structure may have been used for grape crushing.

Belt 2 (intermediate): Three distinct sectors:

1. A 30-m-long rectangular hall connected to a corridor ending at a central shaft. Defended by a lost millstone door, the passage includes compartments resembling animal shelters.
2. Two large rooms, repurposed by locals for storage, contain silos and a well. They connect to the central shaft.
3. Apart from Sector 2, two additional chambers, protected by former millstone doors, connect to the central shaft.

Belt 3 (deep): A steep 70-meter corridor leads to a room on the fourth level. This passage was originally narrow, designed to hinder attackers. Three more rooms, each containing wells, link to the central shaft. Final chamber, at approximately −55 m, is often misidentified as a church. However, based on observations, it likely does not exceed −40 m, aligning with Derinkuyu 4.

Demir theorizes that excavation began with a deep shaft reaching the aquifer, ensuring water supply, ventilation, and debris removal. Corridors and rooms were subsequently dug radially, expanding vertically as well.

Observations indicate that the tourist area contains only four millstone doors, though operational rooms suggest more were once in place. The settlement follows a defensive layout with three bands:

- **First band:** Near the surface, with camouflaged entrances in homes. Tunnels extend downward, protected by millstone doors.
- **Second band:** Provides mobility, with fewer closing mechanisms.
- **Third band:** The final refuge, accessed via a steep, heavily defended tunnel.

This defensive pattern is also evident in other systems, suggesting a coordinated construction effort. A similar principle appears in underground shelters, which are primarily horizontal rather than vertical but also feature a secure inner refuge. This indicates a regional approach to planning and organization (Figs. 4.18 and 4.19).

Fig. 4.18 Derinkuyu Underground Shelter (photo M. Gülyaz)

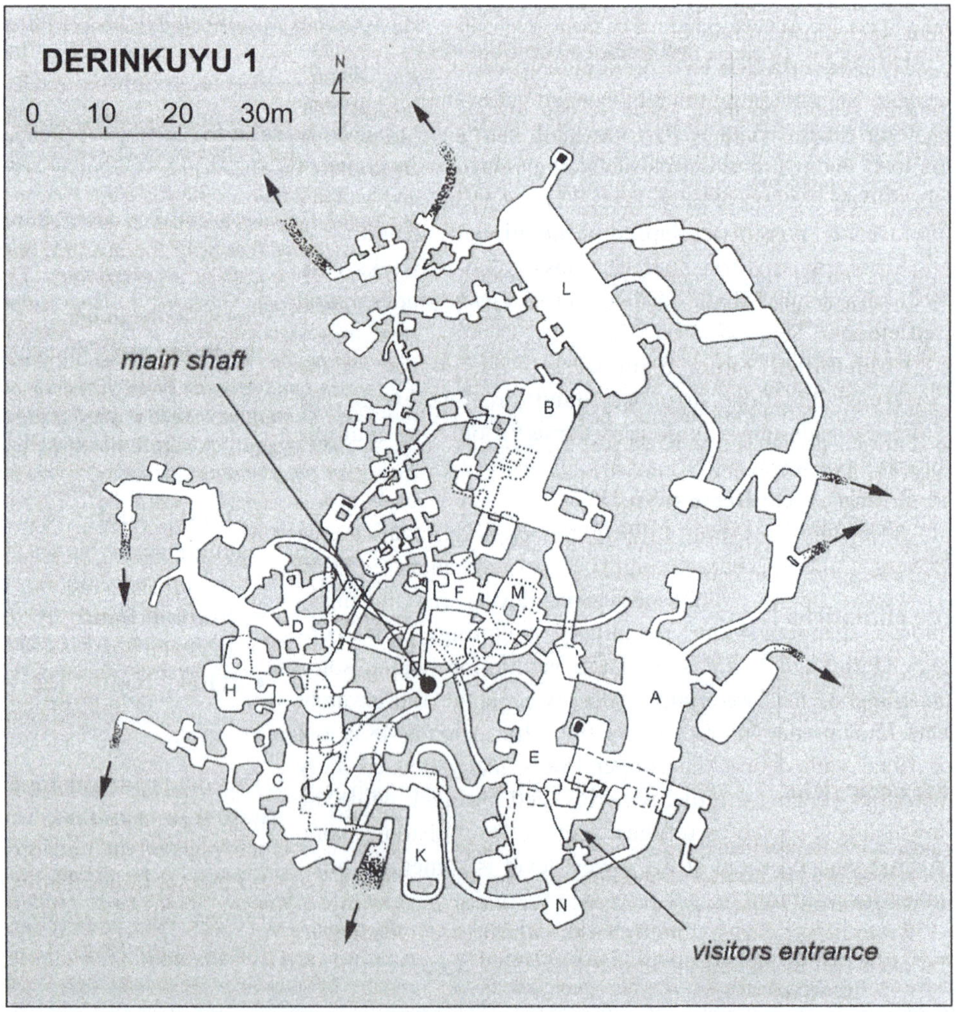

Fig. 4.19 Derinkuyu Underground Shelter sketch plan (Bixio, 2012)

4.14 Doğala Underground Shelter

Doğala Village is located on the slope of Mount Erdas, 9 km northeast of Derinkuyu District. Near the village, the caravanserai ruins belonging to the Seljuk period are found. The first unit called DO1 by Bixio is located on the slope descending from the ruins to the flat valley. DO2 is located 120 m up and is near the fortress ruin on top of the hill. The rough road extending toward the summit of Mount Erdas passes near these two underground shelters.

System DO1: DO1 consists of four interconnected rooms with niches in the walls and service structures. Hall [2] has a well for water, room [3] contains dry storage spaces, and room [4] has a bell-shaped underground deposit, possibly once accompanied by another now-filled deposit. The entrances are protected by millstone doors (M1, M2), with another (M3) securing the tunnel between rooms [1] and [4]. A tunnel from room [4] leads to a central area [5], secured by two additional millstone doors (M4, M5).

Two routes extend from room [5]. The first is a 40-m uphill tunnel, now blocked [6]. The second is a descending shaft leading to a short passage with a missing millstone door (M6). Room [7], long and narrow, contains a flooded descending tunnel on its right wall. In the opposite corner, a hallway, possibly preceded by another operating room, connects to room [8], from which another flooded tunnel begins.

Nearby, fifty meters to the right of the main system, there is a separate group of compartments [9, 10, 11]. Room [9] leads down to hall [11], where two tunnels—one ascending, one descending—are now blocked (Fig. 4.20).

System DO2: Located higher up, DO2 has an entrance near a spring emerging from a blocked passage. A narrow opening leads to a room, from which a tunnel extends downward, also flooded. To the left, another descending corridor ends after a short double route. Unlike DO1, DO2 appears to have been designed to capture groundwater rather than serve as a shelter.

DO1, originally larger, likely functioned as an underground refuge. The current entrances, protected by millstone doors M1 and M2, were probably once preceded by unprotected outer rooms that have since collapsed.

Urban does not determine the age of these structures but notes that the higher terrace contains both fortification ruins and abundant obsidian shards. Oral sources suggest more fortifications exist in the surrounding mountains. Urban also notes that within a 5 km radius of Doğala, he identified six fortifications linked to underground structures, including Çakıllı, Suvermez, and Kaz Kale. Additionally, about 10 km away is the Hittite stele of Topada, dedicated to the great king of Tabal, near the underground settlement of Topada. These findings indicate that the region was a well-fortified and strategically significant area in ancient times (Bixio, 2012).

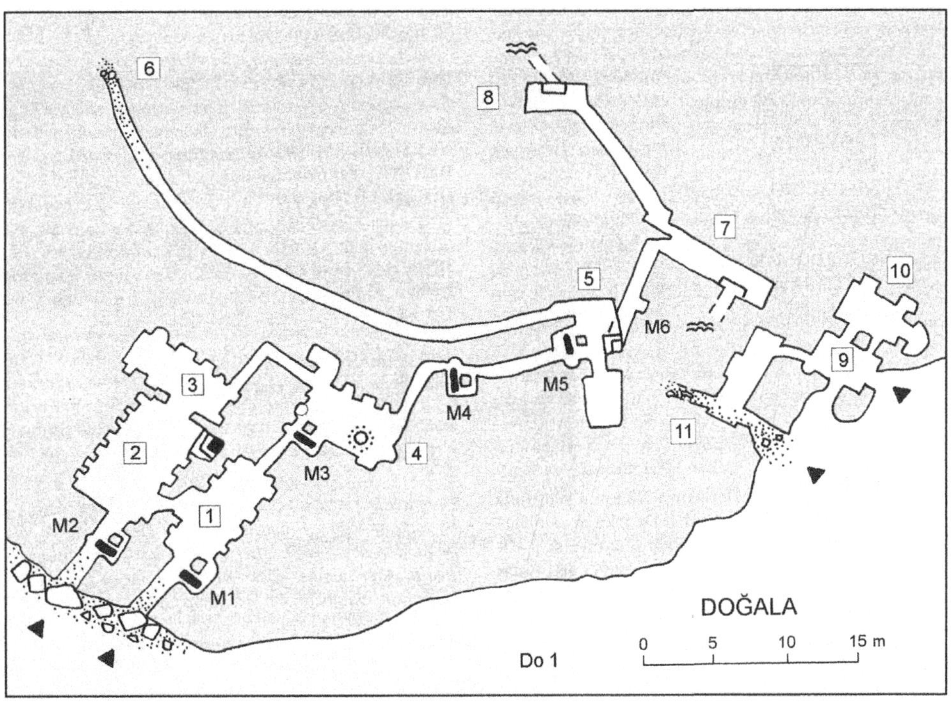

Fig. 4.20 Doğala Underground Shelter System DO1 plan (Bixio, 2012)

4.15 Doğanlı Underground Shelter

Doğanlı Village is located 90 km south of Kayseri, on the border of Niğde province. Even though this region is part of the ancient Cappadocia, this village is distant from all significant settlements. Mentioned previously in different sources (Bixio, 2012; Gülyaz & Yenipınar, 2007; Yörükoğlu et al., 1988), Doğanlı Underground Shelter has been examined in detail, surveyed, and mapped by OBRUK Cave Research Group for the first time (Yamaç & Tok, 2016).

One of the initial architectural features that sets the Doğanlı Underground Shelter apart from other defensive constructions is its location approximately 2 km away from the present-day settlement and on an uninhabited plateau. The presence of ancient cave dwellings in the vicinity of Doğanlı Village provides clear evidence that it was an established settlement in antiquity. If this is indeed the case, it would have been challenging for residents to seek shelter in a defensive structure 2 km away from the settlement during an attack. The most plausible explanation for the peculiar positioning of the Doğanlı Underground Shelter is arguably the plateau on which it was constructed—the site of an ancient settlement, the remains of which are no longer extant. This question can only be answered through comprehensive area surveys and archaeological research.

The main chamber of the Doğanlı Underground Shelter presents a further architectural anomaly. The main chamber of this underground structure, reached after ascending the stairs at the main entrance, is impressive. The stairs, carved from rocks with a slope of 38°, are locally known as "Forty Steps." The chamber, which has a floor area of 500 m^2, is supported by 15 columns and is considerably larger than any other underground chambers of a similar nature (Fig. 4.21). To illustrate, the two largest underground chambers in Cappadocia, which are believed to have been used as churches, are Kemer Ören (400 m^2) and the Church of Derinkuyu (300 m^2). The purpose of construction of the two main chambers of Tatlarin Underground Shelter remains a matter of contention. However, the plan indicates that each chamber is approximately 140 m^2, suggesting that they may have been used for a variety of purposes (Bixio, 2012).

In addition to its considerable dimensions, the architectural details of this chamber are also worthy of note. The north and east walls of the chamber feature 14 distinct rock-carved sections. The term "sections" is used to describe these structures rather than "rooms" because they appear to be divisions separated by thin walls. The discovery of a cistern on the east wall of the chamber, which was found to be perfectly clean water at a depth of 2 m and dimensions of 3 × 2 m, provides compelling evidence that this centuries-old structure displays architectural superiority.

The question of whether the main settlement is the current Doğanlı Village or another now-vanished village remains unanswered. The rationale behind the construction of a 500 m^2 chamber supported by 15 columns in an underground structure, carved for defensive purposes in a remote area far from major settlements, is unclear.

The eastern tunnel, which branches off from the chamber in a different direction, is obstructed by debris after a short distance. In contrast, the northern tunnel extends for a total of 74 m, but then becomes obstructed after narrowing. A further sub-branch of the northern tunnel reaches

Fig. 4.21 Doğanlı Underground Shelter main chamber (photo A. E. Keskin)

the upper level and, alongside a fairly large room, connects to the surface. All tunnel connections leading to the surface are protected by millstone doors. Notwithstanding the absence of some millstone doors, the operational rooms and door slots on the tunnel walls provide tangible evidence of their existence.

The third branch, which emerges at the surface, represents the continuation of the tunnel that deviates to the south at the point where the stairs leading to the main entrance are first encountered. The tunnel contains four carved rooms of considerable dimensions, which differ from those observed on the chamber walls. The surface connection of this branch is via the collapsed ceiling of one of the smaller rooms. Notwithstanding the evidence indicating the existence of an additional entrance in the final chamber, the circumstances precluded the possibility of conducting further investigation due to structural instability. To the east of the aforementioned room is a diminutive chapel, measuring 3 × 2 m. Given the high probability that it was never reused since its construction, there are numerous architecturally and archaeologically unanswered questions pertaining to this remarkable subterranean complex. In our estimation, the lack of answers to these questions, which can only be resolved through an excavation and cleaning, renders the mysteries of this isolated settlement forever unsolved (Fig. 4.22).

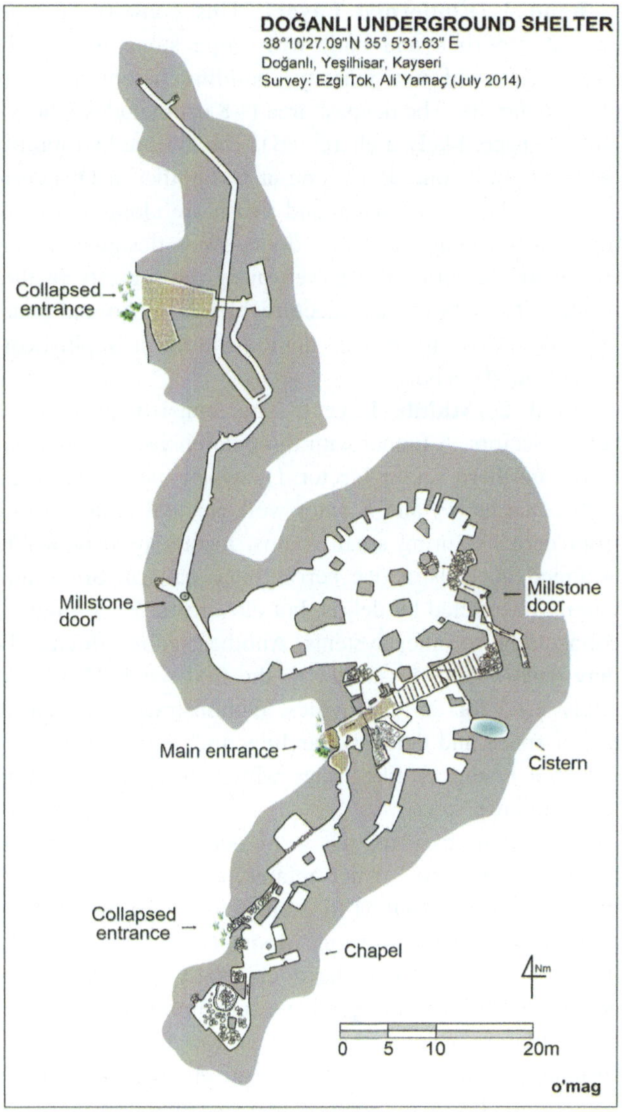

Fig. 4.22 Doğanlı Underground Shelter plan (drawn E. Tok, A. Yamaç)

4.16 Durmuş Evi (Derinkuyu 4) Underground Shelter

Durmuş Evi is an explored underground system near Derinkuyu 1. It has lost its connection to Derinkuyu 1. The entrance of "Durmus Evi" is in the basement of the house located 100 m north of touristic Derinkuyu Underground Shelter. The structure was researched and mapped by the cavers who are the members of the National Commission for Artificial Cavities of the Italian Speleological Society. It is very close to Derinkuyu Underground Shelter and it was connected to this underground shelter until 1991. However, the connections were gradually sealed or intentionally blocked. It is highly possible that the blocked tunnel on the second floor is connected to the section which is open for tourism. The excavations carried out in order to expose the walls of the new structures block the access to the deeper levels. In the first section, the rooms, some of which are partially carved into rock and some erected with large square stone blocks. The entrance on the surface is built around the courtyard and is located in the area where the house belonging to the Durmuş family.

The millstone door found in Area C probably protected the second entrance. There are three different millstone doors extending along the tunnel reaching to the north section. In the north section, there is a colonnaded central hall. A well was found in this section. A few tunnels originate from this point. There are small rooms used as stable and feed mangers carved into rock and holes on both sides. The plan of the second section is almost similar to Derinkuyu 1 and it shows more than one architectural design. The south section consists of two rooms. The room M is encircled by small rooms. It is defined as "winery" since the traces of the production are still visible.

A blocked tunnel on the second level (−19 m, point O) may have connected it to Derinkuyu 1's now-closed access points. Observations describe the structure before disruptions caused by construction in 1999, which blocked access to deeper levels.

Band 1 (Epidermal Layer): This consists of rooms carved into rock and built with large stone blocks. The entrance is within a masonry building belonging to the Durmuş family. The deepest area (−8 m) includes a heavily altered space, likely a church (B). The system had multiple defensive millstone doors: one at C, another at D securing access to the second band, and two more along a descending tunnel (point E). These doors were designed to protect against external threats, ensuring the security of those inside. The rock-cut architecture blends functional spaces with defensive measures, demonstrating a sophisticated underground design.

Band 2 (Middle Layer): This consists of two connected sectors. A tunnel with three defensive closures leads to the northern sector (sector 1), which has a central hall with columns (F) and a water well. Several tunnels extend from here, featuring small rooms, mostly used as stables, arranged in a distinctive herringbone pattern. Some areas remain obstructed by debris, but closures at F1–F4 indicate connections to other systems. Another sealing device (F5) may have linked this sector to the first band. This sector, likely used for daily activities, demonstrates the strategic use of space and defensive architecture, ensuring access to essential resources like water while allowing for autonomous defense.

The southern sector has two rooms, including the so-called "wine room," which retains traces of production. A tunnel connects room M to room N (−19 m), protected by multiple defensive millstone doors (N1, N2). A blocked tunnel at point O may have originally provided further access. A shaft (N3) might have functioned as an entrance or ventilation system, though it is now obstructed. These structural elements indicate a well-planned underground community, balancing security and sustainability.

Band 3 (Deep Layer): A short pit from room N leads to a steeply descending passage ending in a small blind room at −44 m (P), suggesting excavation was halted due to flooding. Along the passage, two operating rooms provided defensive measures, with an unfinished third room further down. This structure mirrors Derinkuyu 1's layout, featuring three overlapping bands: an epidermal surface layer, a complex middle band, and a deep tunnel leading to a last-resort refuge. However, Durmuş Evi lacks completed refuge rooms due to geological conditions. This suggests the excavation was abandoned before reaching its full defensive potential.

A shaft at N3 may have served ventilation and debris removal functions, similar to the central shaft in Derinkuyu 1. The blocked tunnel at O likely connected to another radially developed system, resembling Derinkuyu 1's layout. The depth and complexity of this system highlight its role as both a shelter and a strategic defensive structure, offering insight into the engineering capabilities of its creators.

In total, 10 millstone doors and at least 18 operating rooms have been identified in Durmuş Evi, underscoring its extensive planning and defensive architecture. Despite modern disruptions, the site remains a fascinating example of underground construction and ancient ingenuity, reflecting a highly organized, secure, and self-sufficient shelter (Fig. 4.23) (Bixio, 2012).

4.17 Emiruşağı Underground Shelter

The Emiruşağı Underground Shelter is situated in the village currently known as Emiruşağı, previously referred to as Avşar Obası. This village is located approximately 22 km to the east of the district of Tomarza, which lies within the borders of Kayseri Province. In past years, there were efforts to develop this underground shelter into a tourist attraction. These efforts included the installation of some electrical infrastructure intended to illuminate the interior for visitors. However, these plans appear to have been discontinued, most likely due to a lack of sustained interest or investment.

The architectural structure of the underground complex is composed of two principal galleries that extend along a north–south axis. These galleries are arranged around a central large hall, which spans a diameter of 33 m. This hall is positioned on the hillside of the village, making it a prominent feature within the terrain. Entry to this main hall is facilitated by a primary entrance that has been constructed using stonework. Over time, this entrance has undergone certain alterations, likely as part of the initial attempts to adapt the space for tourism. It is believed that the walls of this entrance were modified during this period. However, it is also highly probable that the ceiling of the entrance collapsed at some point, either due to natural erosion or lack of maintenance.

One of the two main galleries extends toward the south and is currently filled with a significant accumulation of debris and rubble. As the gallery progresses further, it becomes noticeably narrower and subsequently turns in an easterly direction. From this point, it branches out into four separate paths. Unfortunately, all four of these branches are obstructed by deposits of debris, rendering further passage inaccessible. Along these tunnels, which continue toward the south and southeast, only a limited number of rooms have been identified. Additionally, there is a notable absence of typical defensive features commonly found in other similar shelters. Apart from a single millstone door—which has since been displaced from its original location—there are no remaining examples of operational rooms or intact defensive doors in these southern tunnels.

The second gallery extends northward from the central hall and appears to be longer than the gallery directed

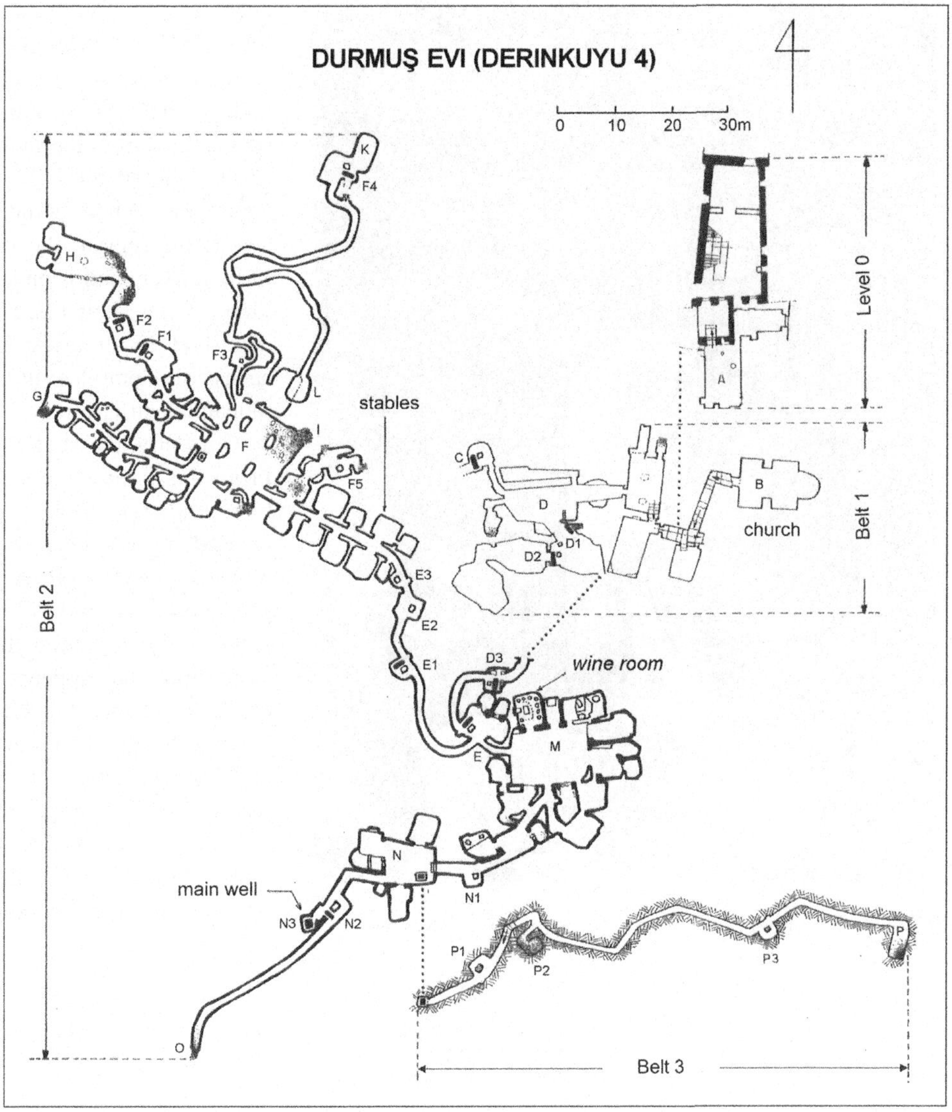

Fig. 4.23 Durmuş Evi Underground Shelter plan (Bixio, 2012)

south. After a distance of approximately 13 m, this gallery diverges into two separate paths near a hall that contains underground storage areas. One of these diverging tunnels turns eastward and continues for about 38 m before ending in a small chamber measuring 16 m^2. Given its isolated location and the length of the approach tunnel, this chamber is presumed to have served as a "Last Defense" room. Despite this assumption, it is surprising that this section does not contain any evidence of a millstone door or an operational room.

The other tunnel, continuing further north from the same bifurcation point, terminates in a blocked passage. Adjacent to this endpoint is a room that contains three silos. These storage silos suggest the space may have been used for food preservation or supplies, providing support for long-term habitation or defensive retreat (Figs. 4.24 and 4.25) (Yamaç, 2018b).

4.18 Eski Pusatlı Underground Shelter

Located on a flat plain in front of a rocky ridge, approximately 2 km north of Pusatlı Village in the Tomarza district, lies a remarkable underground complex locally referred to as "Eski Pusatlı." This hypogenic settlement stands out in the region due to its unique method of construction, which deviates from the more common practice of excavating directly into the rock face. Instead, the inhabitants of Eski Pusatlı dug vertically and horizontally into the flat terrain in front of the rocky wall, forming dwellings beneath the open plain. This distinctive architectural choice sets the site apart from many other examples of underground settlements found throughout the broader region.

Around thirty underground structures have been identified at this location. However, most of them are currently inaccessible due to either partial or complete collapse, or

Fig. 4.24 Emiruşağı Underground Shelter (photo A.E. Keskin)

because they have been filled in with alluvial material over time. These conditions have made detailed investigation of many parts of the settlement difficult. Despite this, one underground shelter has survived in relatively good condition. This particular shelter stands out not only for its preservation but also for the remarkable quality of its excavation technique. When compared to other defensive structures across the region, the craftsmanship and precision visible in its tunnels and rooms are clearly superior. Although portions of the shelter are damaged, collapsed, or have been disturbed by treasure hunters, the sections that remain accessible demonstrate an advanced level of skill in subterranean construction.

Currently, the shelter has only one operational entrance. From this point, a tunnel extends westward and is secured by a millstone door, suggesting it may lead to an exit beyond the chamber that is presently blocked. This main tunnel connects to three large rooms, which are themselves protected by two additional millstone doors. Interestingly, just three meters after the tunnel entrance lies a fixed millstone door that appears to lack an operational room. This detail is notable, as such doors typically feature an adjacent chamber from which the mechanism could be activated.

Another section of the same tunnel veers to the south, leading into a large chamber measuring approximately 45 m^2. This room, like several others, is currently blocked with debris. The lack of any defensive installations or millstone doors in this southern extension implies that this area

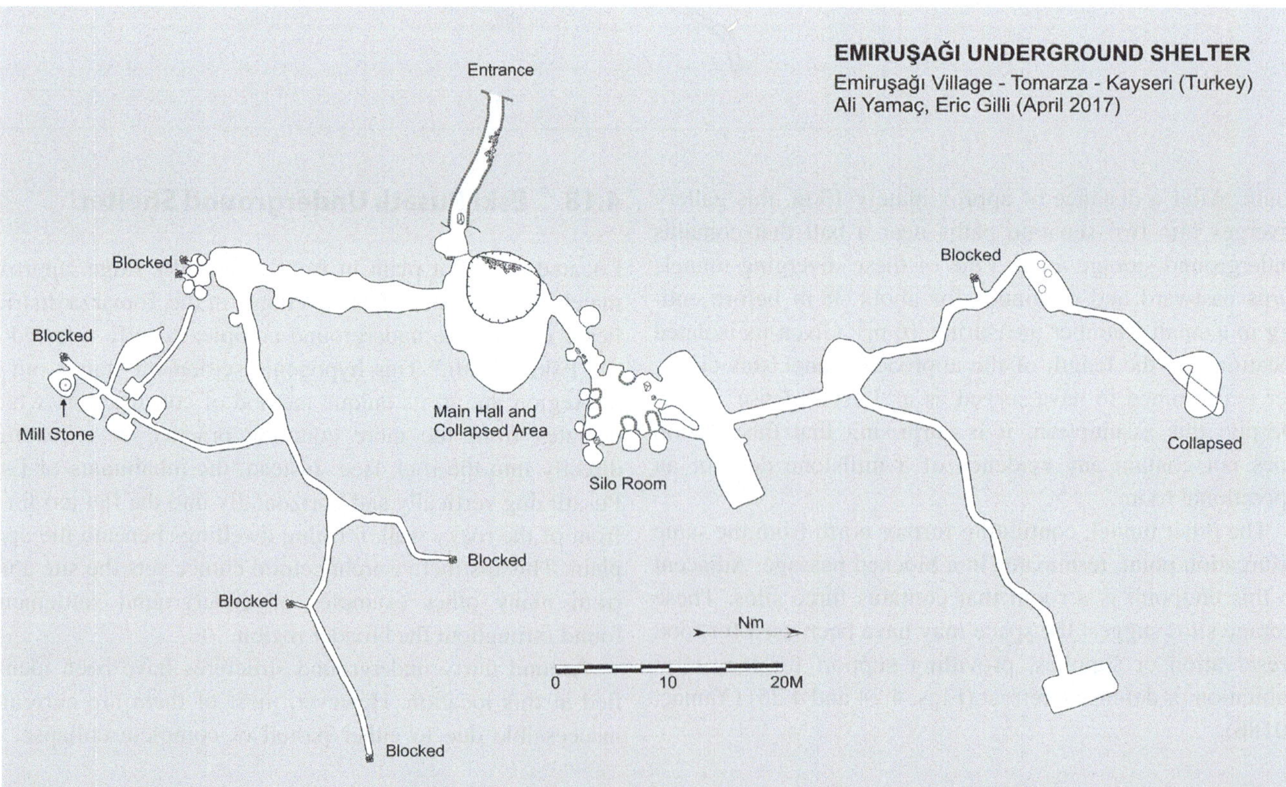

Fig. 4.25 Emiruşağı Underground Shelter plan (drawn A. Yamaç, E. Gilli)

Fig. 4.26 Entrance of Eski Pusatlı Underground Shelter (photo A.E. Keskin)

was likely intended to be an internal part of the shelter rather than a protected boundary or exit.

Further extending northward, the structure ascends through a narrow, inclined tunnel. At the top, after passing through another millstone door, one arrives at an upper level of the complex. This floor includes a total of four entrances—one still open to the surface and three that have become obstructed over time. Structurally, the rooms on this upper floor do not resemble typical defensive shelters and likely represent a later addition. These chambers were, in fact, used as a barn until fairly recently, suggesting an adaptive reuse over time.

Additionally, there are three interconnected rooms on the lower level, located to the east of the entrance tunnel. Despite its name and known features, the underground settlement known as "Old Pusatlı Village" has never been the subject of a formal, scientific archaeological investigation. It is quite likely that further exploration could reveal additional defensive structures in the surrounding area that have yet to be uncovered or documented (Figs. 4.26 and 4.27) (Yamaç, 2018b).

4.19 Fatinler Underground Shelter

The Fatinler Underground Shelter is a well-planned underground defense structure located approximately 500 m east of Şıh Barak Village in the Tomarza district of Kayseri. It is situated on a slope that overlooks a wide valley, providing a strategic vantage point. The shelter features a total of three separate entrances, each positioned to serve a specific function within the structure's overall defensive design.

One of the main entrances is situated on the north wall of a large underground barn, which is characterized by walls that contain small niches carved into the surface. This entrance leads into a tunnel that is protected by two consecutive millstone doors, placed in such a way as to increase the security and defensive capability of the shelter from any approach through the northern section. These doors likely played a significant role in slowing down intruders and providing enough time for internal defense mechanisms to be activated.

The eastern portion of the structure is built at a slightly higher elevation compared to the rest of the complex. During our examination of the site, it was discovered that there are three different tunnels in this section that extend eastward. However, all three of these tunnels have been blocked by debris and rubble, making it difficult to determine their full length and structure. If these tunnels do not descend to a lower level and instead continue directly toward the valley floor, it becomes clear that they cannot extend very far, given the elevation and the surrounding landscape.

A second entrance to the shelter is located in the central part of the structure. This entrance is also protected by a millstone door and has its own operational room adjacent to it, likely used to monitor movement and control access through this part of the complex. A notable architectural feature of the shelter is the presence of another millstone door located right in the middle of the structure. This door was strategically placed to guard against potential attacks coming from the south and to protect the three rooms located in the northern section of the shelter.

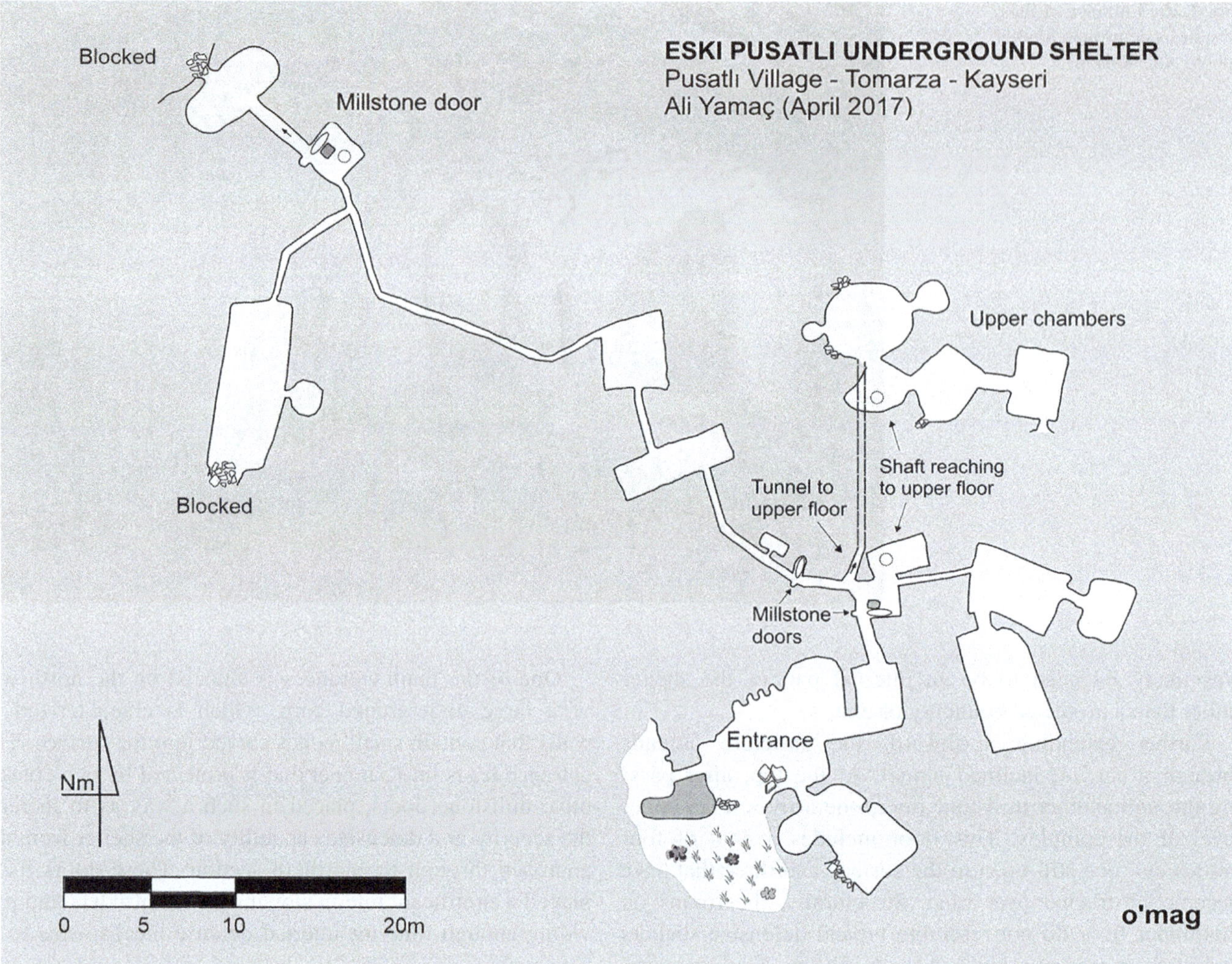

Fig. 4.27 Eski Pusatlı Underground Shelter (drawn A. Yamaç)

One of the most remarkable and architecturally significant aspects of the Fatinler Underground Shelter is the high number of millstone doors present within such a compact structure. Despite having a main axis measuring only 80 m and comprising seven rooms in total, the shelter contains five millstone doors—four of which remain in situ. The defensive orientation and placement of these doors suggest that the structure could be divided into at least three independent and self-contained sections, allowing for localized defense in case of a breach in any part of the shelter.

The tunnel located at the southernmost end of the shelter ascends slightly as it approaches the surface, and, like the others, it is blocked with rubble. Given its close proximity to the surface and its sloping upward direction, it is assumed that this tunnel, too, does not continue for a long distance. This conclusion aligns with observations made in similar blocked tunnels throughout the site. (Figs. 4.28 and 4.29) (Yamaç, 2018b).

4.20 Filiktepe Underground Shelter

From the Acigöl district, the road that turns north on the main road to the west of Nevşehir reaches the village of Ovaören after 26 km. The underground structure develops within the butte of Filiktepe (also known as Filikören), which is situated 1.5 km southeast of this village in the vast plain. The National Commission for Artificial Cavities of the Italian Speleological Society's members conducted research and surveys on this sanctuary between 1999 and 2002.

The underground shelter's entrances are situated to the south, southeast, and east of the hill's wall. However, the existence of an additional and extensive underground system that was excavated at a lower level is attested to by a number of craters that are nearly entirely submerged at the base of the butte.

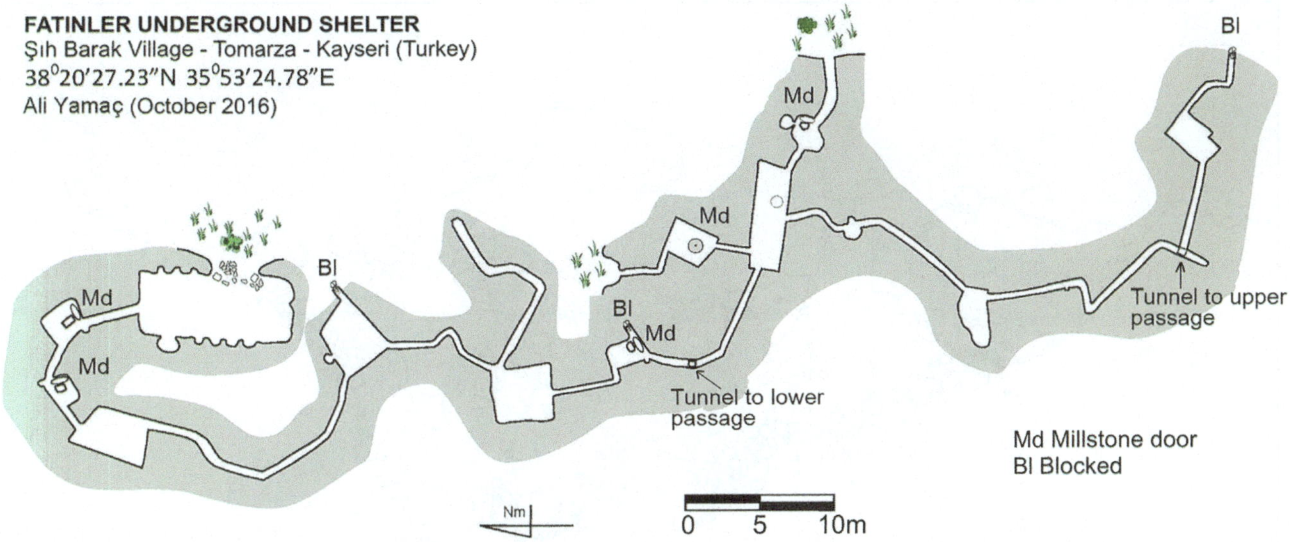

Fig. 4.28 Fatinler Underground Shelter (drawn A. Yamaç)

Fig. 4.29 Fatinler Underground Shelter (photo A.E. Keskin)

The system that has been investigated is of unexpected complexity and scale, as its entrances are visible in the vertical wall of the butte. Nine substantial systems, some of which are interconnected, were identified during the investigation. Based on the most recent data, six subsystems are identified, each of which comprises independent chambers.

- System F1/F2/F3 (approximately 638 m)
  A complex underground system is the result of the connection of the two subsystems F1/F2 and F3 by a long and tormented inner tunnel, which is reached through a succession of entrances to the southern end of the butte. Each one is furnished with a well and numerous millstone doors that are repeatedly fired. A false pit in F2 leads to several chambers below, which are protected by millstone doors and a lower access that is currently filled with earth. The F3 system is evidently distinguished by an abundance of chambers that are likely utilized for storage. Clear evidence of both effractions and defensive emergency measures has been discovered in the latter system (Fig. 4.30).
- System F4 (approximately 105 m)
  The portion that was investigated replicates the straightforward design of a "redoubt" that is equipped with an

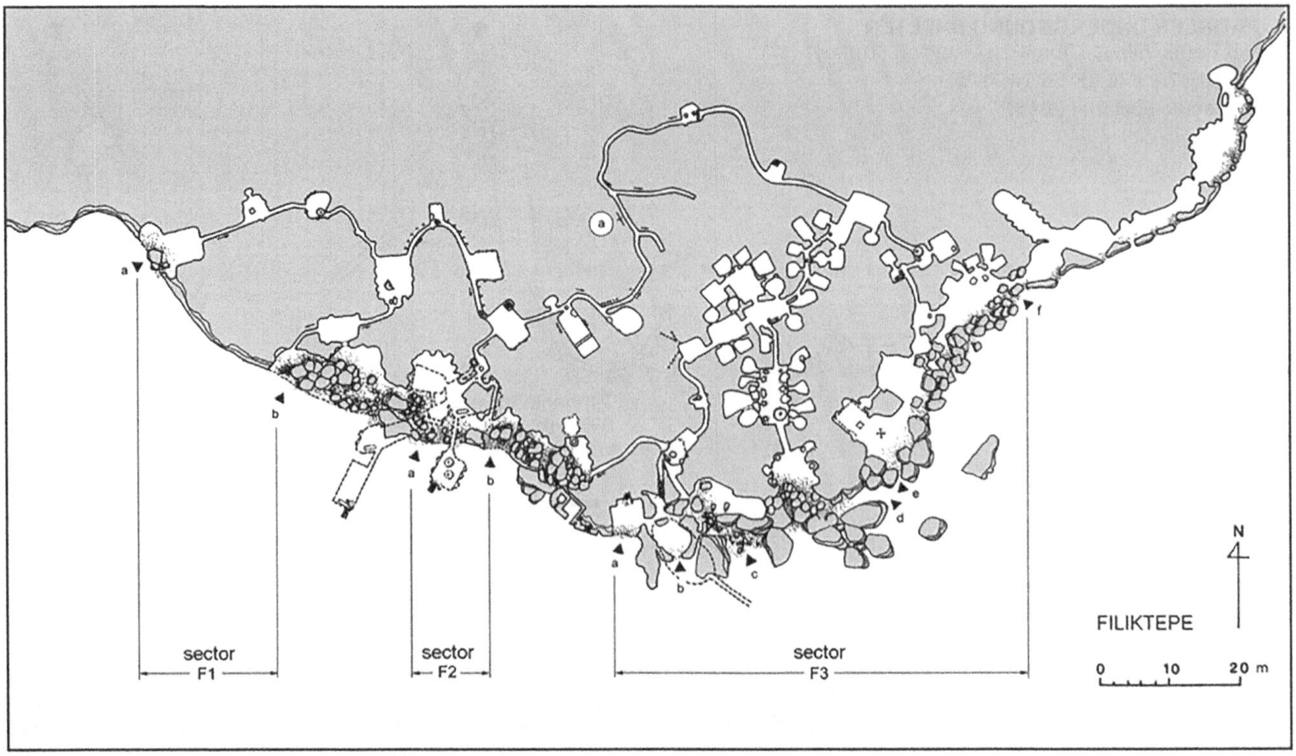

Fig. 4.30 Filiktepe Underground Shelter, plan of sectors F1, F2 and F3 (Bixio, 2012)

emergency tunnel (an "escape route"). This scheme, which is also present in the adjacent butte of Ovaören, is slightly more complex in this instance. The "redoubt" is not immediately defended at its entrance; rather, it is protected by a millstone door situated in a separate operating room along the tunnel. Subsequently, the chamber is linked to an even more intimate room. It is also important to consider the inner branch of the "escape way," which is safeguarded by an additional millstone door that leads to the interior. This door may be used to establish a connection with the extremely near system F5.

- System F5 (approximately 327 m)
  It is composed of four contiguous "redoubts" that are connected by tunnels and are interspersed with additional interior chambers. The presence of a double millstone door to obstruct the passage is particularly noteworthy in two "redoubts." The unfinished passage is also noteworthy, as it may have been intended for the connection to system F4, which also contains a well, similar to system F3.
- System F6 (approximately 150 m)
  Three "redoubts" are installed in the chambers that are in direct communication with the exterior, two of which are interconnected.
- System F7 and F8 (approximately 72 m)
  It is a matter of chambers that are abundantly filled with earth and are difficult to interpret.
- System F9 (approximately 381 m)
  It comprises four "redoubts," one of which provides simultaneous access to two outdoor areas (F9/d, F9/e), which are connected by additional indoor chambers. It is important to reiterate the existence of a well in the lengthy connecting conduit. It is observed that the residential areas' entrances are situated to the south and east, while the northern portion of the hill is home to only a few chamber-like monuments (Fig. 4.31).

Several large rooms have been excavated in the first band of the cliff, which is closest to the outside. Many of these rooms are concealed or camouflaged by extensive rock wall collapses. Some of them appear to be capable of identifying churches or chapels, as if each of the systems described above was centered around distinct family groups, each with its own place of devotion. Tunnels are excavated into the inner band to connect individual subterranean shelters, thereby facilitating a more effective defense against enemy attacks.

The following closing devices were identified within the investigated net, whether or not they were equipped with millstone doors:

12 in the system F2-F3; 10 in the system F3; 3 in the system F4; 9 in the system F5; 4 in the system F6; 1 in the system F8; and 15 in the system F9. A total of 55 devices.

The existence of Christian symbols (crosses in relief) and locations that are likely to have been used for devotion

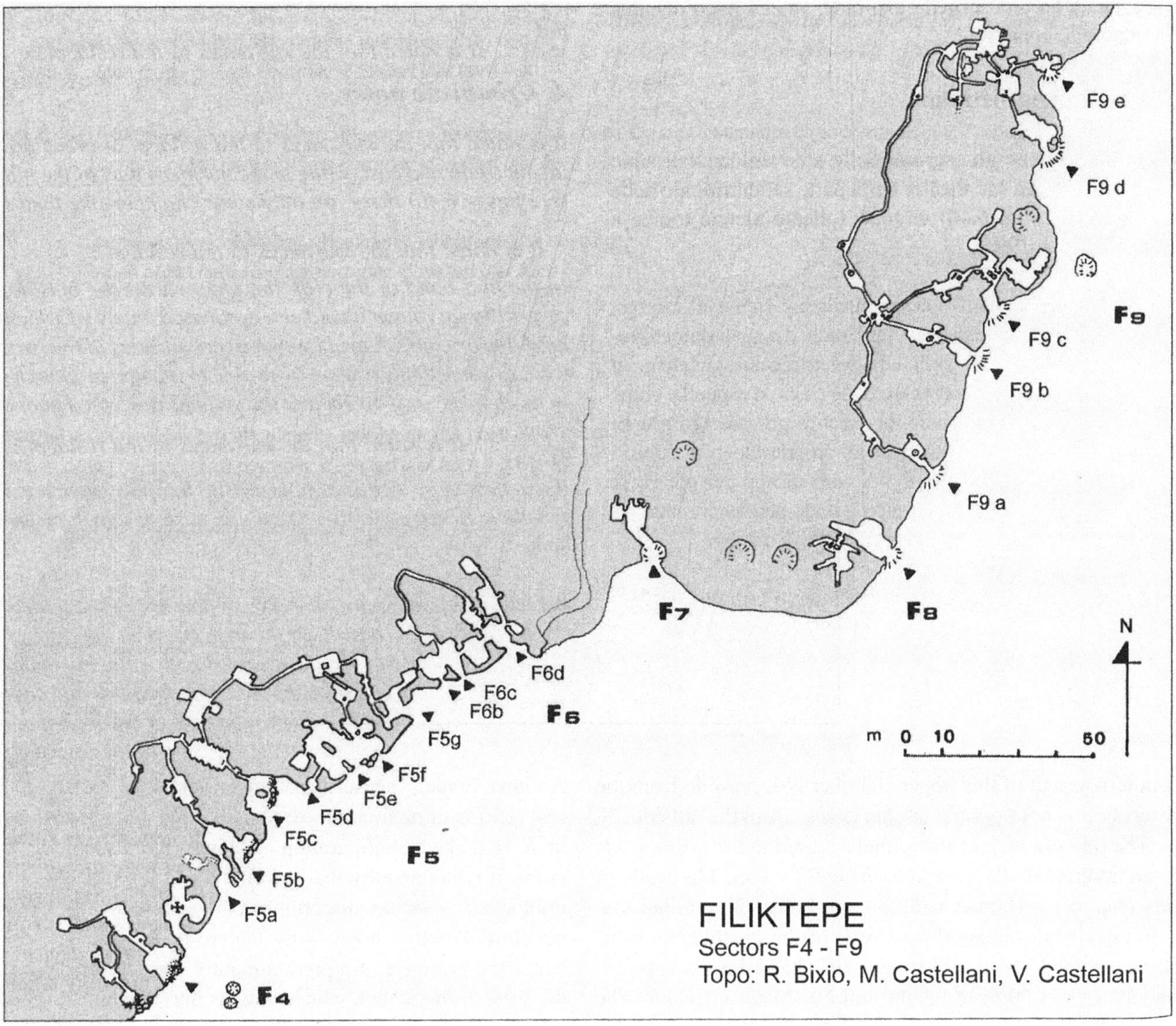

Fig. 4.31 Filiktepe Underground Shelter, plan of sectors F4–F9 (Bixio, 2012)

indicate that the site was in use during the Byzantine period. Pottery fragments that protrude from the ground in front are also Byzantine. It is important to note that the apse in the chamber adjacent to the entrance of F3/e is interrupted by a tunnel, which would logically indicate that it was constructed after the church's use. Additionally, the area between F4 and F5 may serve as a church, as it features a niche that resembles a pit-like grave.

4.21 Golgoli Underground Shelter No 1

The village of Mustafapasa (Sinassos) is situated 17 km to the south of the Ürgüp District of Nevşehir. An ancient village of cone dwellings is located in the area, known as either Golgoli or Gorgoli, in close proximity to Ayvalı Village, situated 5 km to the south of Sinassos. In this settlement, which was abandoned many years ago, there are tens of houses, two churches, and two underground shelters of differing construction. Despite the absence of documented historical or archaeological evidence, this settlement may represent the initial settlement of Sinasos (Kıvılcım, 2011).

The principal entrance to the Golgoli No 1 Underground Shelter is located at the base of the eastern wall of a substantial rock-dwelling structure, situated to the north of the hillside where the primary cave dwellings of Golgoli are located. Immediately adjacent to the entrance is the first millstone door. Subsequently, the primary tunnel is bifurcated into two distinct sections (Fig. 4.32).

Upon reaching the western end of the passageway, one encounters an upper chamber. Despite the fact that the entire

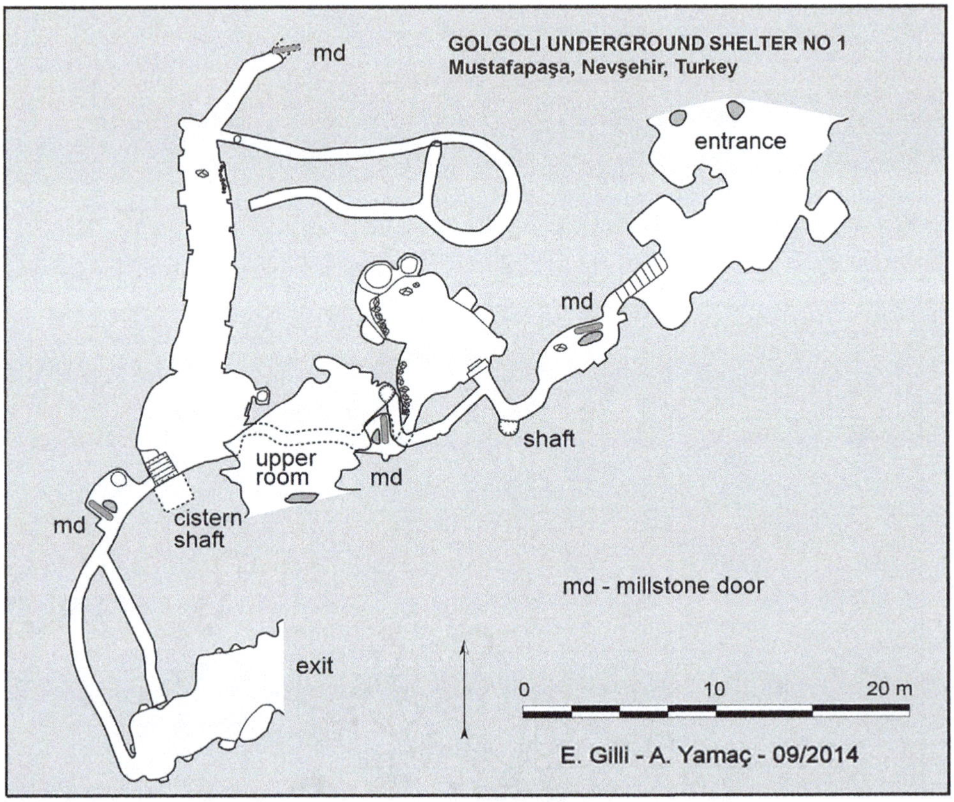

Fig. 4.32 Golgoli Underground Shelter No 1 (drawn E. Gilli, A. Yamaç)

southern aspect of this upper chamber is accessible from the exterior, it is not possible to gain access from the outside.

The other branch of the tunnel, located to the north, leads to an additional chamber measuring 9 by 6 m. The walls of this chamber are lined with storage holes. The tunnel that continues from the southwest wall of the chamber is fortified by an additional millstone door. The ceiling is approximately 1.40–1.60 m in height and 60–70 cm in width. The tunnel then reaches another chamber of notable interest. The initial section is 5 m in width, subsequently narrowing to a 3-m section comprising multiple boulders and measuring 17 m in length. Additionally, stone rings carved into the ignimbrite have been identified, yet their purpose remains uncertain. An unfinished millstone door is situated in close proximity to the rock formation at the point where the tunnel reaches the chamber. It seems likely that the carving was halted due to the cessation of the threat of raids, resulting in the abandonment of this defensive structure.

It is possible that other factors may have contributed to the cessation of construction, such as the threat of invasion or the emergence of a local disease, or a lack of funding prior to the completion of the project. This example illustrates the location and the method of carving of the numerous millstone doors that can be observed in Cappadocian underground shelters.

A tunnel extending from the northern extremity of the chamber is equipped with a millstone door, while the subsequent tunnel is obstructed by accumulated debris (Fig. 4.33). Another tunnel, measuring 32 m in length, is located to the east of the aforementioned tunnel. This tunnel terminates in a U-shaped configuration, lacking any discernible endpoint. It is noteworthy that the entirety of the tunnel remains unblocked, with no discernible connection to any external structure. Another noteworthy feature of the primary chamber is the presence of a predominantly vertical pit situated to the west of the tunnel, which extends into this tunnel. A stairway, measuring 80 cm in width and with a slope of 650 mm, descends to a depth of 7.5 m and terminates at a cistern or well. Despite the absence of water in the cistern (or well) at the present time, it may be the same structure referred to by the local population of Sinasos as the "sacred spring water." The cross carved on the rock at the western wall of the cistern serves to emphasize the religious importance of this building. It may also be posited that this pit functions as a type of defense trap for the tunnel that continues through the south of the stairs. This tunnel, following another millstone door and operation room, reaches a cave dwelling house that opens to Golgoli Valley (Yamaç & Gilli, 2016).

4.22 Golgoli Underground Shelter No 2

Unlike the previous structure, this one is located on the top of a hill, near the Golgotha cave dwellings. This three-floor underground structure is accessed through an opening in the western wall of a geological formation known as the "fairy

Fig. 4.33 Millstone door in the upper room of Golgoli Underground Shelter No 1 (photo E. Gilli)

chimney." In Cappadocia, in contrast to the more gradual ascents observed in other shelters, there are also examples of excavated ascents that are entirely vertical. These are found in conical remnants also known as "fairy chimneys."

An example of a rare defensive structure is the Golgoli Underground Shelter No 2. Another such structure is located in the upper Zemi Valley (Göreme), to the northeast of Karabulut Church (Bobrovskyy et al., 2015). These complexes are excavated with a predominantly vertical development of rooms (overlapping) and are grouped around an axis that is more or less central. This enables the excavated spaces within the rock mass of a single pinnacle to be maintained while the sequence of defense-closing devices remains consistent with the broader strategy of the defensive system.

The southern entrance of Golgoli Underground Shelter No 2 begins with a narrow tunnel that leads to a large chamber protected by a millstone door, situated approximately 7 m away. Another tunnel, situated to the west of this chamber, is characterized by a low ceiling. After a 1-m-high shaft, the tunnel connects to another chamber measuring 4 × 2 m. A third tunnel that continues to the north of the aforementioned chamber reaches the northern chamber. Another millstone door is located within this chamber in its original position. This chamber provides access to an upper level tunnel. The medium level comprises a double tunnel and a room. The staircase connecting the medium story to the upper and lower levels is almost shaft-like in shape, with a height of 6.60 m. The millstone door in the third story, accessible via the aforementioned stairs, is situated at the front and is notable for its considerable dimensions, measuring 1.70 m in diameter and 32 cm in thickness. Such a large and substantial door is highly uncommon in the region. It can be reasonably deduced that the uppermost chamber at the top of this fairy chimney must be the last defense point (Fig. 4.34) (Yamaç & Gilli, 2016).

4.23 Gomeda Underground Shelter

Gomeda Underground Shelter, located in the Gomeda Valley near Mustafapasa, is a well-preserved cave complex. It is part of a larger rock settlement carved into a cliffside, featuring residential and utility rooms, rock churches, and defensive shelters with wells and millstone doors.

At the top of the cliff, remnants of a large rock complex include a well-preserved cave church. The church's entrance is in a partially destroyed gallery-like room. The church itself has a complex layout, suggesting multiple construction stages. Initially, it had a basilica layout with a transverse single-aisle naos, three apses, an altar niche, and rock benches. Later, small single-aisle churches were added in place of the side apses, each with horseshoe-shaped apses, wall thrones, altar niches, and seating for priests. Additional altar niches were added in the third stage, dating the church's development between the tenth and thirteenth centuries.

The entrance room originally functioned as a narthex. The church complex was later expanded with additional rooms: Room 3, likely a monastic cell with niche-shelves and a sleeping area; Room 4, a Christian tomb with burial pits and niches; and Room 5, another monastic cell with sleeping niches and storage pits. A water well (K1) in Room 5 may have originally been a cistern. Over time, cave shelters expanded from Room 5, including hidden chambers and defensive structures.

One such shelter evolved from connected utility pits, creating a concealed cave tier (Room 6). A manhole in Room 6 led to Room 7, which opened to the cliff face. Another shelter, based around Room 10, was accessed via Well K1 and a steeply inclined gallery (11). Room 10 had a rectangular layout with niches and an opening in the wall of Well K1, suggesting it originally had a defensive feature. Tool marks indicate Room 10 was developed from Gallery 11, which was constructed later for additional access.

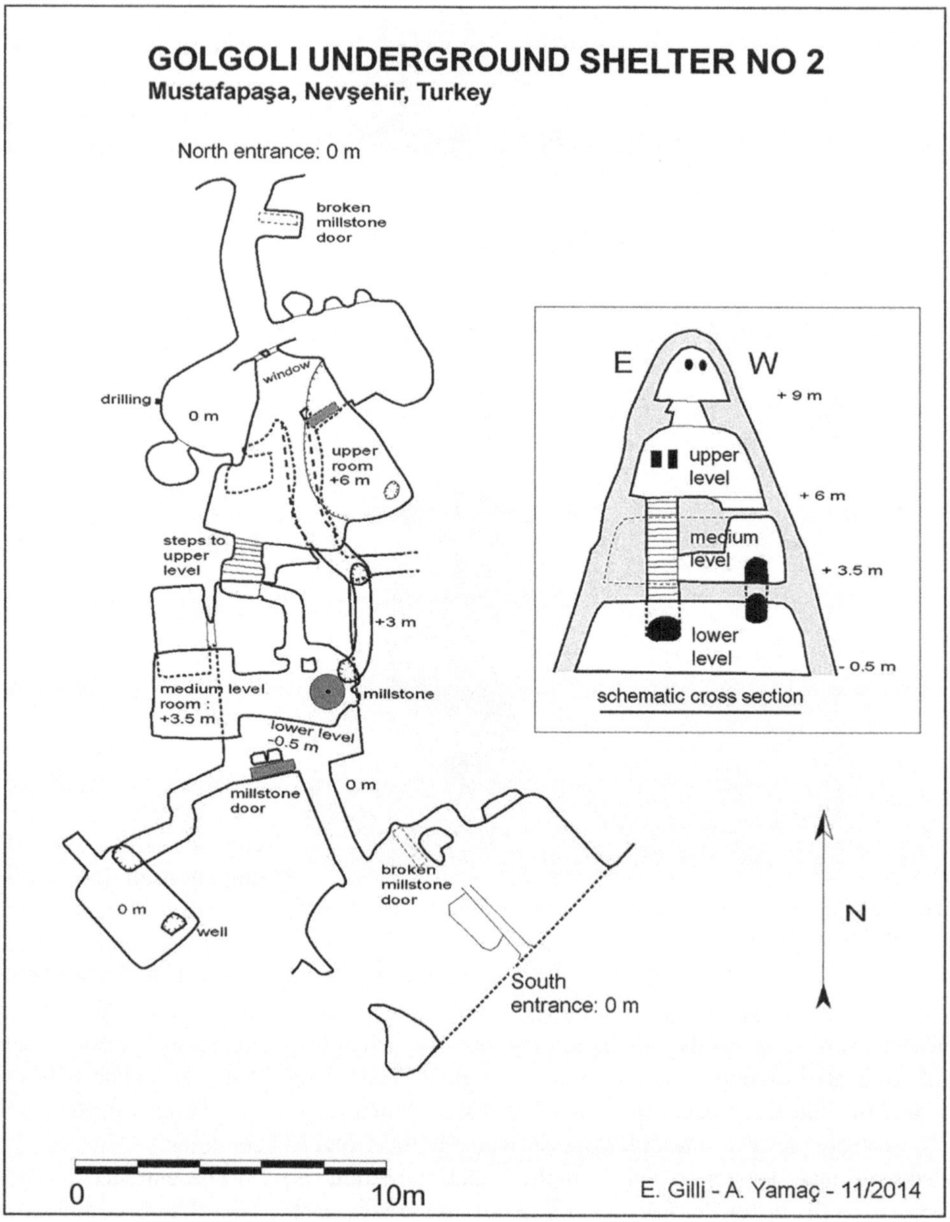

Fig. 4.34 Golgoli Underground Shelter No 2 plan (drawn E. Gilli, A. Yamaç)

From Room 10, Passage 10b led to chambers 12 and 13. Chamber 12 had storage features similar to other Cappadocian shelters, while Chamber 13 initially served as a passage but was later modified with vessel recesses. It ended in a dead-end with small windows and a cave toilet, similar to other underground shelters in the region.

Another shelter developed from a grain pit in the southeastern corner of Room 5. A well (K2) was cut into its base, concealed by a slab, leading to a steep gallery (8). This gallery connected to another well (K3) and ultimately to Room 9, a rectangular chamber with niches and a possible cistern. Room 9, the most secure chamber in the shelter, extended into Passage 9a, a dead-end that may have been intended as an evacuation tunnel. Similar tunnels, both completed and unfinished, have been found in other Cappadocian shelters.

Two additional galleries were recorded from Room 10 toward Chamber 9. Gallery 10a connected with Passage 8 at the base of Well K2, while Gallery 10v linked Gallery 10b and Well K3. These passages formed a labyrinthine structure integrating separate shelters (Fig. 4.35).

The Gomeda complex employs techniques common in Cappadocian shelters, such as converting storage pits into shelters (as seen in Mazı), camouflaging chamber openings within water wells (like in Kemerkaya and Derinkuyu), and using vertical communication wells (found in Mazı, Meskendir Valley, and Göreme). The complex evolved in

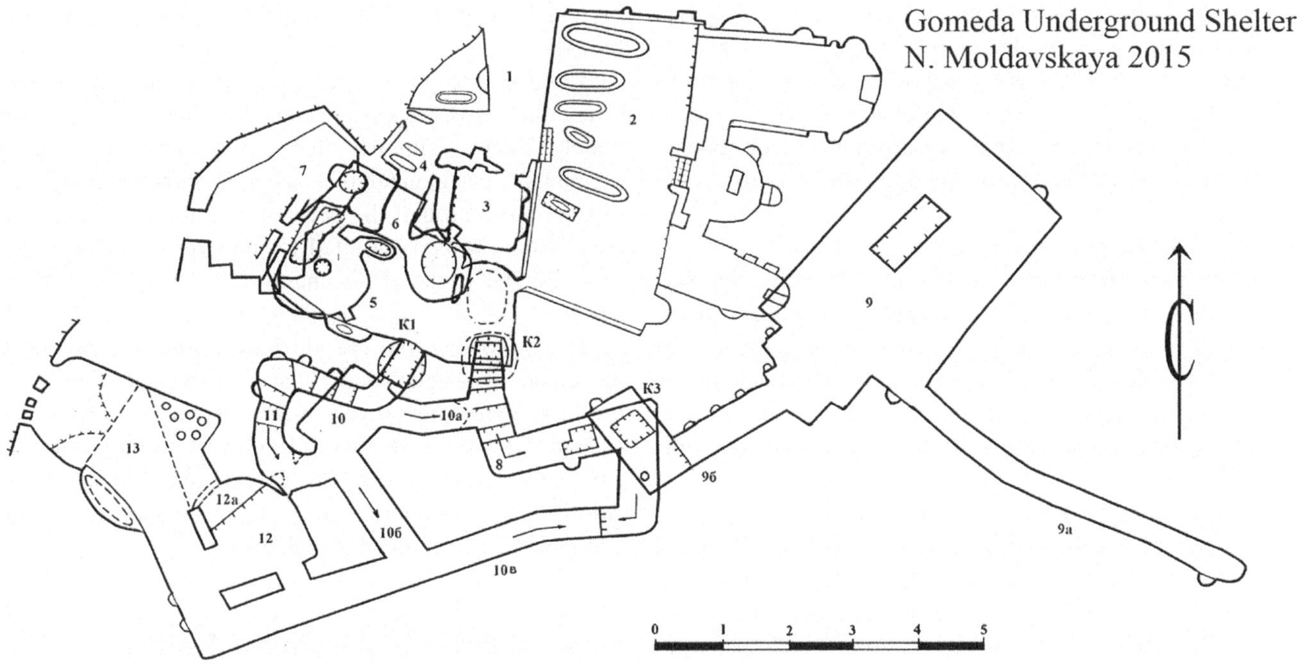

Fig. 4.35 Gomeda Underground Shelter plan (drawn N. Moldavskaya)

multiple stages, initially as separate shelters that were later interconnected. The final phase formed a continuous labyrinth with vertical and horizontal expansion.

Unlike many underground cities, this shelter combined features of both rock settlements and subterranean complexes. While some parts opened onto the cliff face, much of the shelter extended deep into the rock. This suggests an evolution from a rock church and residential complex into a fortified underground refuge.

The shelter was built after the church complex, which dates between the tenth and fourteenth centuries, indicating a medieval origin. This complex stands as an example of how Cappadocian cave dwellings evolved over time, blending habitation, religious, and defensive functions into a sophisticated underground system (Bobrovsky et al., 2019).

4.24 Göstesin (Ovaören) Underground Shelter

The stabilized road turns to the north when it is reached to Acigöl District on the motorway in the west of Nevşehir Province. After 26 km, this stabilized road reaches Ovaören Village. This underground shelter (also called Ovaören) is grouped around the southern end of a small elliptical flat rock (butte) that rises about ten meters over the surrounding plain. Among the houses of the village remains of semi-underground structures emerge, set up on arc structures formed by squared blocks of stone, covered from the outside with earth filling. Along the butte wall, behind the village, both on right and left side of the houses in surface, there are many openings from which one can access the underground structures. It was researched and mapped by the members of the National Commission for Artificial Cavities of the Italian Speleological Society, in 1999–2002.

Five distinct underground systems have been identified during the investigations, each of which comprises multiple chambers that are mutually connected. These rooms, in turn, constitute the subsystems:

- System G1 (approximately 258 m).
  G1/a and G1/c are two entrances on the west side of the butte, while G1/b is an entrance on the east side. These entrances provide access to the broader underground rooms (redoubt), which are typically safeguarded by millstone doors that remain in place.
  The redoubt G1/a is equipped with a well and is linked to G1/b, which is situated on the opposite side of the butte, by a conduit that traverses the entire body of the rocky hill. Some chambers of modest size are visible along this tunnel, which are likely utilized for the storage of supplies. The G1/a area is connected to the apartments in G1/c by an additional tunnel that is perpendicular to the previous one. This passage extends beyond the redoubt G1/c until it is intercepted by the earth. It is likely that it originally reached the surface.
- System G2 (approximately 254 m).
  It is composed of two subsystems that are situated on opposing sides of the butte and are connected by a lengthy internal corridor (linking passage). In close

analogy to the areas G1/a and G1/b of the previous system G1, the redoubt subsystem G2/a contains a well and is connected to G2/b on the opposite side by a long tunnel that likely features storage chambers. It is evident in both subsystems that an emergency tunnel (escape route) exists, which directly connects the redoubts to the exterior.

- System G3 (approximately 127 m).
  At the northern end of the east wall of the butte, by-pass tunnels that were excavated behind the structures themselves connect several chambers that have been significantly altered but still appear to be partly natural. A millstone door that appears to have been abandoned during the transport process is located in the connecting tunnel between G3/a and G3/b.
- System G4 (approximately 112 m).
  The entrance is situated on the eastern side of the butte, in a position that is halfway between G2 and G3. The portion of the system that can be investigated is comprised of a series of chambers from which passages originate, which are presently earth-filled and have the potential to access the innermost systems.
  A tunnel commences from the final room (redoubt), which is safeguarded by a millstone door. After a few meters, dried stones partially obstruct the tunnel voluntarily. It appears to be oriented toward the exterior; this is likely the escape route that is included in nearly all underground shelters in Cappadocia.
  A flight of stairs ascends from the "redoubt" to intercept a partially explorable system that has been excavated at a higher level. A long and serpentine horizontal tunnel, which is perpendicular to the connecting light, extends in both directions. The tunnel continues within the hill on one side until it reaches a large room from which two branches emerge, which are promptly obstructed. It undoubtedly reached the exterior in the opposite direction; the entrance is currently covered in earth.
- System G5 (approximately 105 m).
  It is situated at the southernmost point of the butte. It is composed of two contiguous subsystems, each with its own entrance, denoted by the abbreviations G5/a, G5/b, and G5/c. These subsystems are connected by short tunnels. The G5/a system provides access to another system that has been excavated at a lower level and has only been partially investigated. It is equipped with a multifaceted defense system that includes three millstone doors that are oriented outward to safeguard two parallel entrances that are presently earth-filled.

In conclusion, the settlement of Göstesin appears to contain a comprehensive collection of subterranean shelters that have been excavated on both the east and west sides of the butte. The most significant of these shelters are connected by lengthy corridors that span from one side of the hill to the other.

The fact that the tunnels were excavated at the same time, likely from opposite ends and meeting at approximately halfway, implies that the community would be organized in a manner that separate families gravitate to distinct areas around the perimeter of the hill. This would enable them to be connected for underground defense purposes in the event of an enemy attack, with coordinated action.

The following closure devices (with or without the respective millstone door) have been identified within the investigated net: seven in the system G1, seven in the system G2, two in the system G3, two in the system G4, and eight in the system G5. Twenty-six devices in total. Additional investigations may be necessary to discover additional information (Fig. 4.36).

4.25 Güzelöz Underground Shelter

The Güzelöz Shelter is situated on the northwestern slope of the ancient settlement of Güzelöz Village, which is affiliated with the Kayseri-Yeşilhisar district. It is located between 80 and 100 m from the village. The depression at this point is not the original entrance to this underground shelter. It is clear that the main entrance to this structure is from the tunnel on the back wall of a rock-cut complex located on the rock wall approximately 150 m above this depression, which is defended by a millstone door.

From an architectural perspective, the double tunnel located just east of the entrance to the lower depression is particularly noteworthy. Both tunnels are protected by two colossal millstone doors, each measuring 1.90 cm in diameter. The tunnel that continues in a southerly direction, presumably toward the village situated at the foot of the slope, is obstructed after a relatively short distance.

The main tunnel, which extends in a northeast direction from this lower entrance, is notable for its exceptional width and height, which are rarely observed in subterranean urban settlements. This tunnel, which has an average width and height of 4 m and 2 m, respectively, extends for 54 m with a 20° slope and approaches the upper floor. The shaft, which reaches a height of 6.5 m, is located at the end of this lower tunnel (Fig. 4.37).

The upper tunnel, which continues for 47 m with the same slope but narrows at the upper part of the shaft, reaches the chambers dug into the rocky cliff. All the rooms and storages of Güzelöz Underground Shelter are located on this upper floor.

There are seven different chambers on this rocky upper floor, at least five of which are clearly used as storage. The walls of these chambers are full of niches and, in addition,

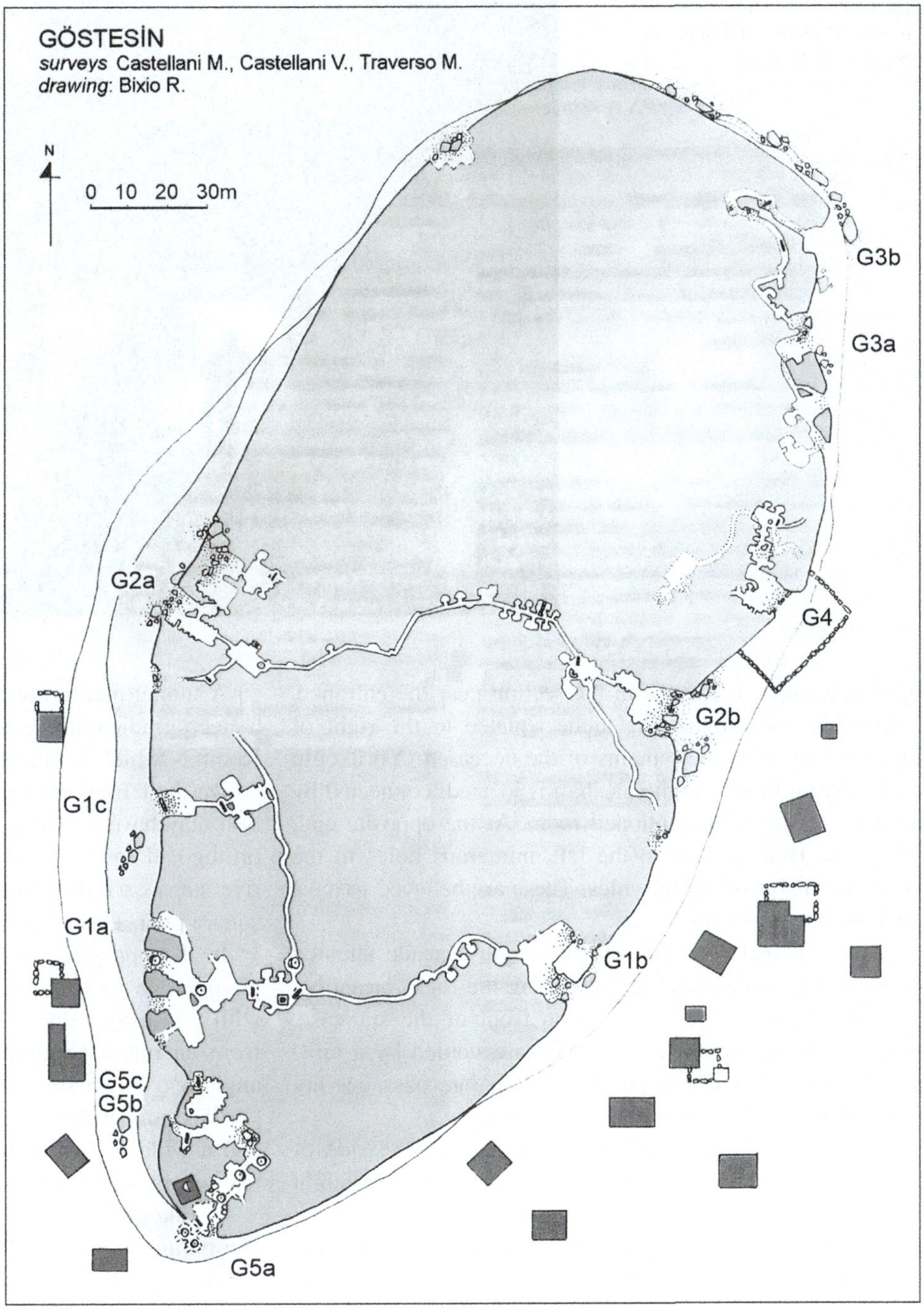

Fig. 4.36 Göstesin (Ovaören) Underground Shelter (Bixio, 2012)

holes were dug in the floors of three chambers to place amphorae. The number and size of the rooms/storages in this upper level, the only entrance of which is protected by a millstone door, are surprising.

Despite all these storages, there is no room that can be used as a residence in the investigated part of the underground city and there is no blocked gallery on the upper level. The only logical explanation for this situation could be that there are living spaces in the continuation of the blocked tunnel at the lower level, which reaches the old village settlements (Fig. 4.38).

4.26 Güzelyurt (Gelveri) Underground Shelter

There are five different underground shelters in Güzelyurt in the southeast of Aksaray Province. The most significant of these is situated on the southeastern slope of Cevizli Street, in close proximity to the center.

The structure is accessed via the main entrance, which opens onto a staircase leading to a hall measuring 30 m in length and 9 m in width. A small room, likely to have been

Fig. 4.37 Main passage of Güzelöz Underground Shelter (photo A. E. Keskin)

used as a toilet, is located in the section near the entrance. Yörükoğlu posited that the niche situated to the right of the stairs contains the remains of the deceased (Yörükoğlu et al., 1988). In the vicinity, a shaft is located, connected by a tunnel to the aforementioned room. At the opposite end, where the room curves to the left, numerous holes in the floor are visible along the sides. These are believed to have been used to house amphoras.

Three distinct passages on the lengthy façade situated opposite the entranceway extend along the rock formation. The initial passageway, situated in front of the staircase, leads to the square room, which is safeguarded by a millstone door. Beyond this point, two separate passages are obstructed by debris after a short distance.

The second passage is located on the opposite side of the courtyard and comprises a tunnel that makes a slight right turn. In the central section, the tunnel ascends beyond a substantial area and subsequently reaches the planned square section. On the right-hand wall is situated an operating room, accessed via a millstone door. This is designed in such a way as to defend the aforementioned area from potential threats originating from the inner sections, which evidently had independent exits to the surface. It is currently not possible to verify this information due to the complete burial of the tunnel in question, along with another tunnel located on the side of the operating room.

Upon returning to the entrance hall, a third access point is encountered, which is formed by a half wall and leads into a room protected by a millstone door via a short sloping tunnel. This defensive mechanism permitted those who sought refuge within the subterranean structure's innermost confines to repel potential external threats.

A tunnel measuring approximately 10 m in length, with a steep gradient and a high ceiling, connects Room 4 with Room 5, which is situated at a lower level.

On the left, a small space (10) of an L-shaped configuration may have served as a toilet. A gallery with a right axis orthogonal to the previous structure descends, comprising five steps carved into the rock, up to a large room with a square plan. This room contains two wells for water supply.

In the opposing corner is an additional structure comprising two rectangular rooms, one of which is equipped with a millstone door (M4) that serves to defend this area from an internal tunnel. The tunnel culminates in a right angle and a few steps, which lead to another square room [7]. The tunnel descends once more, comprising three steps and a wide arc to the right. However, after approximately ten meters, it is obstructed by sediment.

Yörükoğlu (1988) indicates the presence of a conduit for communications carved in the stone [8].

The layout of this settlement appears to adhere to the conventional model of an underground shelter, exhibiting a number of distinctive variations. The initial room, situated in close proximity to the surface, is characterized by a substantial size and is presumed to have been utilized on a continuous basis, including for residential purposes. This is supported by the presence of a well, linked to a "way of escape," a toilet and housing for amphoras, as well as the considerable dimensions of the underground chamber.

In the event of danger, the inhabitants could temporarily take refuge in the inner area (the "redoubt"), which was defended by a millstone door. This consists of a primary space, which contains two wells, ancillary rooms

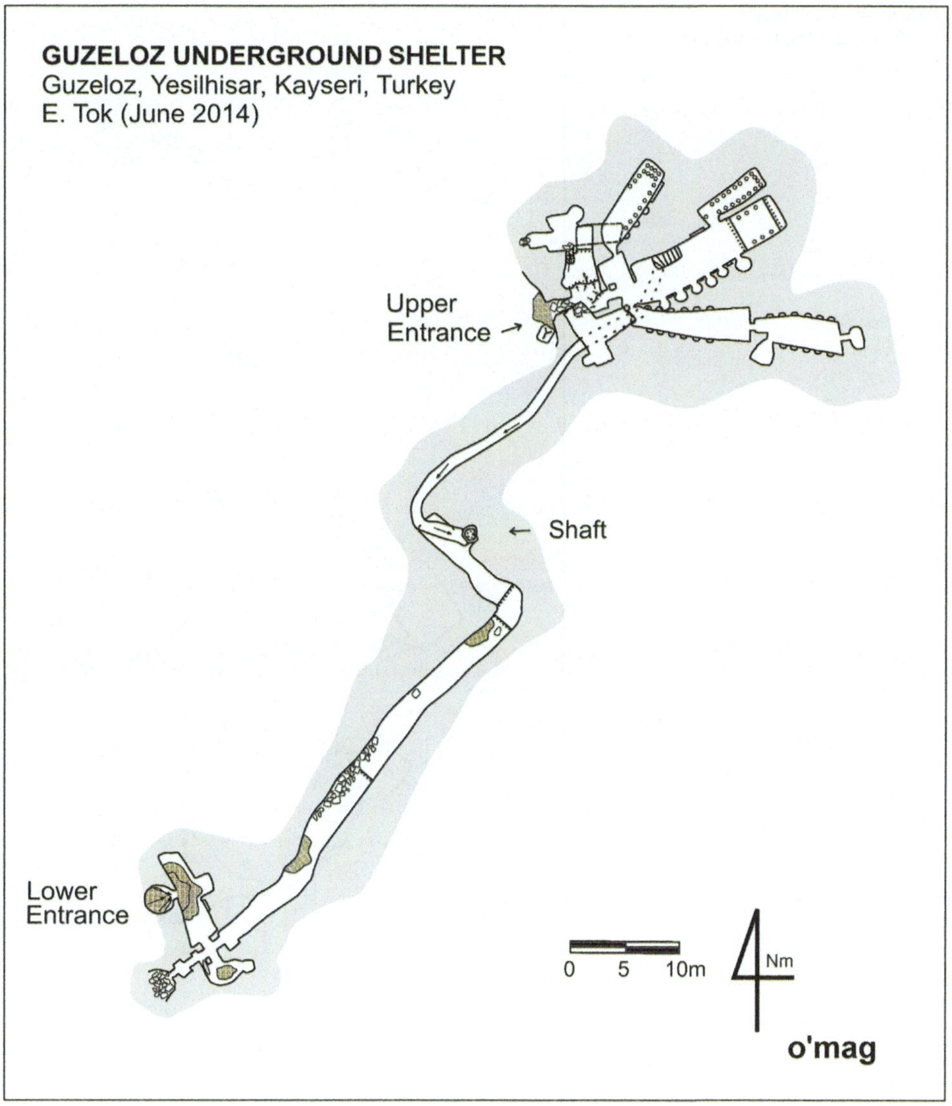

Fig. 4.38 Güzelöz Underground Shelter plan (drawn E. Tok)

(storerooms, toilet facilities), and the typical "escape route" leading to the secondary space.

Currently lacking an exit, the tunnel could serve as an alternative route to the surface, or as a means to access other areas likely defended by a network of communication pipelines, currently an alternative route to the surface. These areas were probably subsequently linked to the residential zone via the room, forming a more expansive and complex settlement.

Both toilets [9] and [10] feature a deep hole and two elevated platforms on their sides. They are located in a small room, set back from the main access corridor, and oriented perpendicularly. Researchers have identified a similar structure in the underground settlement of Tatlarin and Selime. The described underground shelter is just one of several underground structures in and around Güzelyurt. The cliff, above and in front of which the surface buildings of the village extend, houses many rooms excavated at various altitudes, which together can be defined as a "cliff village" (Fig. 4.39).

4.27 Karşı Kilise Underground Shelter No 1

Located 2 km south of the city center of Gülşehir, this structure is a striking example of how a thousand years of historical and cultural value should not be protected. Karşı Kilise Underground Shelter No 1 and No 2 are surrounded by structures built in the last decade.

The rocky settlement extends in the north wall of the cliff and also in the west side of the narrow orthogonal valley coming from the south. The cliff village is composed of a series of superimposed rooms, from the ground level to the top of the cliff, for about 15 m in height, that extend for more than 150 m on the north side. It is evident that the wall of the cliff is in large part collapsed, leaving in sight

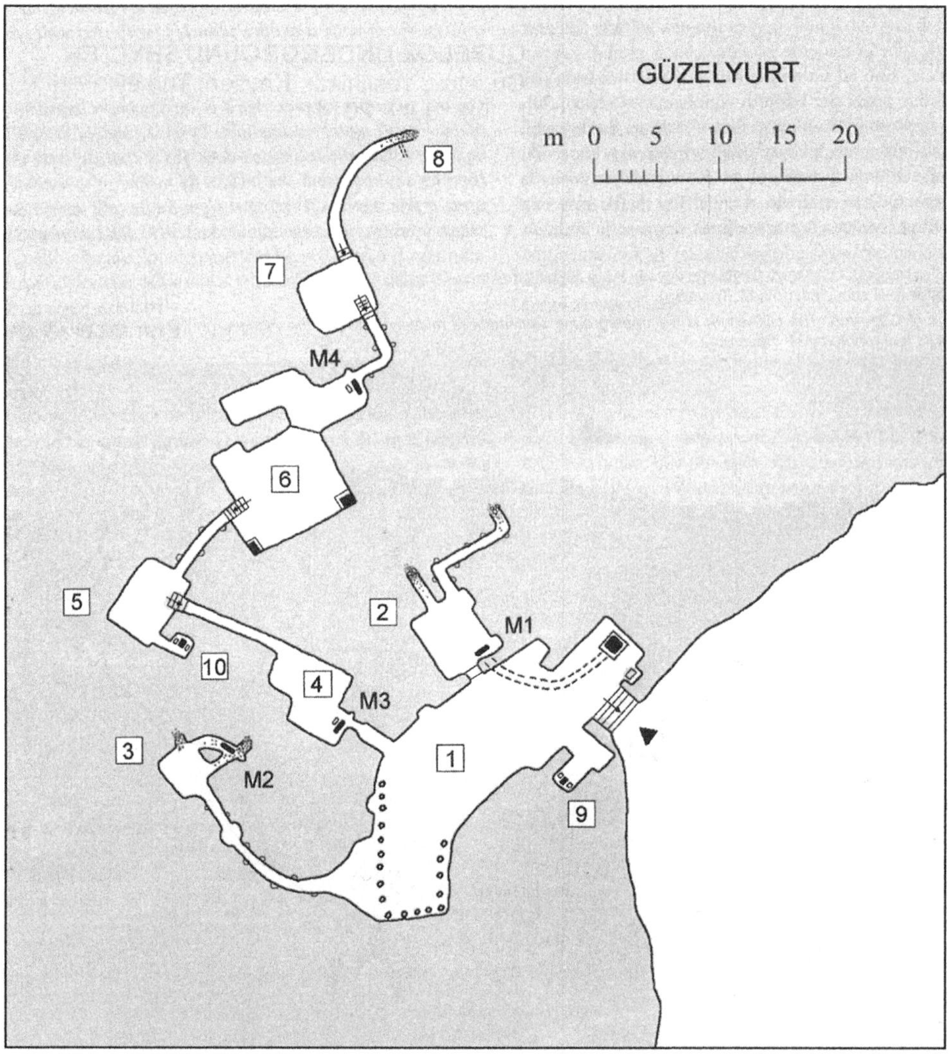

Fig. 4.39 Güzelyurt Underground Shelter plan (Bixio, 2012)

the sectioned rooms. Among them several large pigeon-houses are recognizable, some of them with spheroidal-vaulted niches, others with triangular niches.

There was a tunnel between the village and the church in front of the village (Demir, 1990). The excavated dwellings near the outer wall of the cliff commune with a structure, which functions clearly as an underground refuge, carved into the deepest sections of the rock. One of the rooms in the lower section of the cliff intersects the narrow corridor extending parallel to the cliff and then heading to the rooms in the upper section of the wall. This winding corridor doesn't expand along the road and therefore, only one person can walk within. Toward the end of the tunnel, the corridor gets narrower so that one person can only move by crawling. Then, the tunnel reaches the hall that contains numerous rooms. This long tunnel was reinforced by a defense mechanism. In the entrance, there are traces indicating the existence of a millstone door (M1). Moreover, a hollow (S1) shows that the corridor could have been blocked by a wooden door. The vertical narrow passage (S2) was used for passive defense and it cuts the tunnel. The upper entrance can be closed with a lid placed into a cove. Before the tunnel reaches the region "villagerue," the hollow (S3) in the last section indicates that there was a wooden door in this section as well. The existence of multiple hurdles, the winding and narrow nature of the tunnel show that the tunnel is well protected. In this section of the tunnel, there are other important features as well. In the enlarged section after passing the millstone door, there is a well (P1). In addition to this, the walls in the entrance of the tunnel are decorated with about ten cross motifs. These crosses are placed over or in front of the hollows which were carved for lamps. The area of "corridor-dwelling" consists of a large central boulevard that contains dwelling rooms and these dwelling rooms face a series of wall niches on one side and the silos on the other side. The living rooms of these rooms cover an area of 10 m^2 and the plans of these rooms vary. The small corridor with low ceiling—which is about 1 m high—links each room to the main tunnel. Niches used for lamps were carved into the walls of the tunnel near the entrance of two rooms and each niche is decorated with cross motifs. The first tunnel of two [3] located at the end of the "corridor-dwellings" heads to the rock with no exit. The other one extends along the steep over the stone house [4] and then reaches outside. The millstone door (M2) with a thickness of 29 cm and a diameter of 90 cm at the west end of the corridor-dwellings could have been used to block one of these tunnels. The corridor-dwellings with its halls and the silos was the last refuge (redoubt) of the whole settlement (Fig. 4.40).

The underground shelter, with its long and well-protected access corridor and the corridor-dwelling with its clear defensive role, then had the characteristics of a redoubt (the last refuge) to serve the community that lived in the rocky village. It could accommodate a limited number of refugees when compared to other larger settlements, such as Derinkuyu and Kaymaklı.

The existence of numerous carved crosses is undeniably a sign of Christian frequentation of this site. This might suggest a date between the eighth and tenth centuries, a period during which the Christian community was subjected to incessant Arabic raids. However, the presence in the vicinity of the monastery (also protected by defensive devices), which paintings certainly date back to 1212, may suggest a shelter intended only to protect themselves from the phenomenon of banditry arising in the period of internal conflict of the Seljuk empire during the twelfth and thirteenth centuries.

4.28 Karşı Kilise Underground Shelter No 2

This site is located approximately 2 km south of Gülşehir, on the road toward Nevşehir. A short driveway leads to the entrance of the church, dug at the foot of a group of tufa pinnacles. From the square, continuing a few hundred meters to the west you reach the cliff village described in the previous record. The current Turkish name of this rocky structure, Karşı Kilise, means "the opposite church," evidently due to the fact that it is placed in front of the cliff village Büyük Kale.

The entrance to the rocky church [0] opens in the west wall of a group of pinnacles, straightened and adorned with blind arches, horseshoe-shaped, carved in bas-relief. The bottom of the wall, evidently in danger of collapse, was consolidated with mortar, as well as the top of the pinnacle where stone slabs to prevent infiltration of rain-water inside the church were placed, too. The restoration also included the cleaning and consolidation of the paintings inside and the construction of an exterior wall designed to prevent

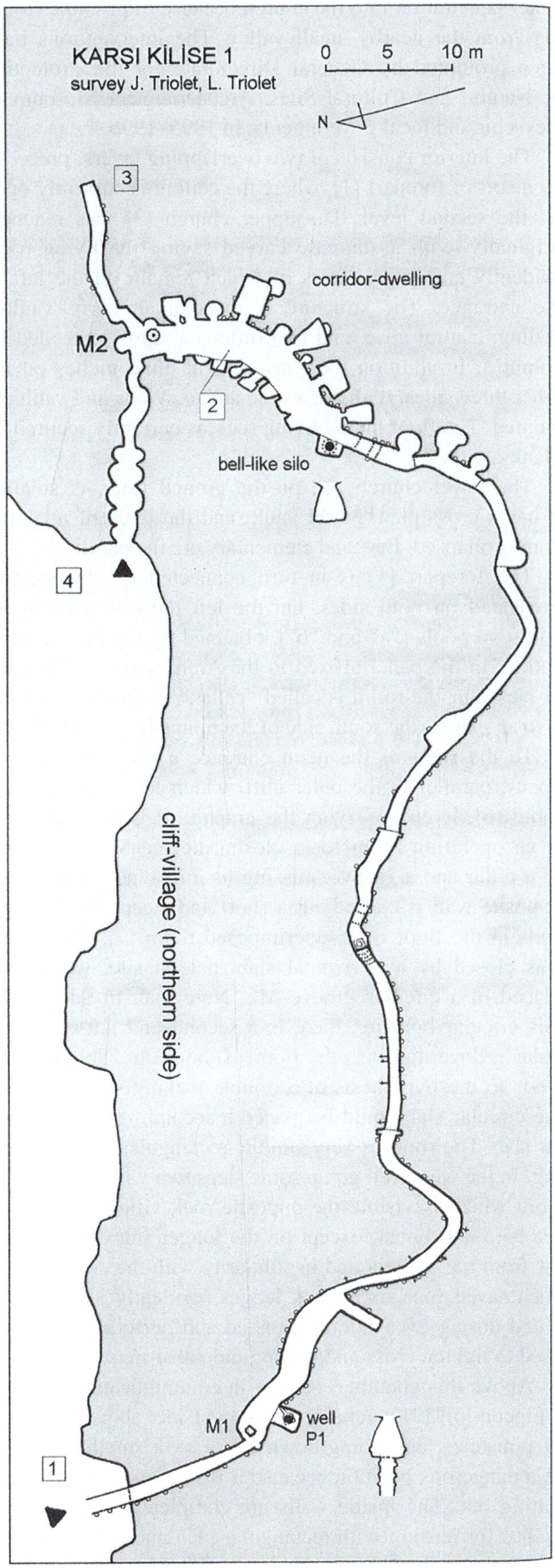

Fig. 4.40 Karşı Kilise Underground Shelter No 1 plan (Bixio, 2012)

water penetration into the church because of possible flooding from the nearby small valley. The interventions have been promoted by General Directorate for the Protection of Natural and Cultural Sites, Archaeological Museum of Nevşehir and local governments, in 1995–1996.

The interior consists of two overlapping layers, preceded by a sort of forepart [1], where the ceiling is currently open on the second level. The upper church [3] was reached, originally, with a staircase carved saving the living rock, evidently collapsed, traces of which remain on the left of the entrance. The structure is rectangular, barrel-vaulted ceiling, central apse, and two little side apses preceded by columns. In addition there are several blind niches on the other three sides, with horseshoe arches. Walls and vault are painted. The floor, in the living rock, is currently secured by planks.

The lower church [2], on the ground floor, is smaller. It has a cross plan, barrel vault, and the trace of a central dome collapsed. Few and elementary are the paintings.

The forepart [1] is in turn connected to other rooms excavated on both sides. On the left there is a room [4] with two pools ["a" and "b"], obtained by saving the excavation. In the panel affixed to the Archaeological Museum of Nevşehir the room is called "cellar." Probably there was also an exit on the south side of the pinnacle (point "c").

To the right of the main entrance a short passage [5] opens, parallel to the outer cliff, which leads into a small room [6]. Even observing the graphic, it is clear that this is an operating room for a closing device M1, consisting of a pillar and a groove: missing its millstone door. On the opposite wall is carved out a short and steep staircase that ends in the floor of a superimposed room [7]. The mouth was closed by a horizontal slab, not on site, which was placed in a circular groove M2. Note that, in addition to this circular housing, there is a second and larger rectangular groove dug into the floor of the room. This element leads to the hypothesis of a double locking system: above the circular slab could be placed a second larger rectangular slab. The room is very simple, rectangular, and flat ceiling. In the west wall go up some steps to a wide window [8] from which is visible the opposite rock village. The walls are bare and linear, except on the longer sides, where, not far from the corners and in solidarity with the ceiling, have been saved four small rock ledges (evidently already provided during excavation), rounded and perforated, perhaps used to tighten cross and/or longitudinal strings.

Above this chamber, but not in communication, there is a pigeon-loft [9] reachable via the outside, above the group of pinnacles, and going down as far as a small side door, in a dangerous position because it is dug into the rock wall jutting out. The inside walls are completely covered with niches for nesting with rectangular plan and arched mouth; they are dug in staggered rows, overlapping, except for the side to the outside which is evidently collapsed probably due to water erosion that excavated the base of the cliff, again established recently with mortar.

We observe that while the church, with its two floors, is almost impressive, work spaces and shelter are rather modest in size. It is possible that the same shelter had also served as a cell for the monks, whose numbers had to be very limited, if not a single one, especially since there are no structures recognizable as cenoby. Keep in mind, however, that the underground portion could be complemented by external buildings or other rooms carved into the rock, long since lost, although it is believed that the church remained in use until the early twentieth century.

It is evident that the small monastery has an underground shelter independent of the cliff village. Despite the small size, it had two closing devices, one vertical (M1 in room 6) and one horizontal (M2 in room 7). In our opinion, the large window [8] was originally closed, or perhaps reduced to a narrow slot to control any raiders. Under present conditions, it would be a weakness on defense. The collapse of a portion of the rock wall, as above described, may actually have produced significant changes even in the rock mass. In addition, 5 m above the ground, below the window [8], the traces of a rectangular excavation can be seen: it could correspond to the end of the tunnel connecting the rocky village, or the beginning of the excavation of a new room, not completed (Fig. 4.41).

4.29 Kaymaklı Underground Shelter

This underground shelter is located in the south of Cappadocia, on the main road that connects Nevşehir to Niğde, about 19 km from the capital of the province. The complex is constructed beneath a hill of tufa in the center of the village of Kaymaklı. The underground shelter has been accessible to tourists since 1968. The opening of the site to the public resulted in structural modifications that fundamentally altered its character.

The network of Kaymaklı presents a significant challenge for interpretation due to the presence of multiple and subsequent excavations, which have introduced considerable complexity. It is practically impossible to provide a detailed account of this network. The construction of additional rooms has resulted in the interruption of several corridors. Some of the original underground stores, which were constructed within the confines of a room, have been exposed and are currently visible in the rooms located on the lower level. Those that remain intact are separated from the next rooms by a screen of rock, just a few inches in thickness.

The occurrence of numerous sinkholes has resulted in the formation of a multitude of inadvertently interconnected

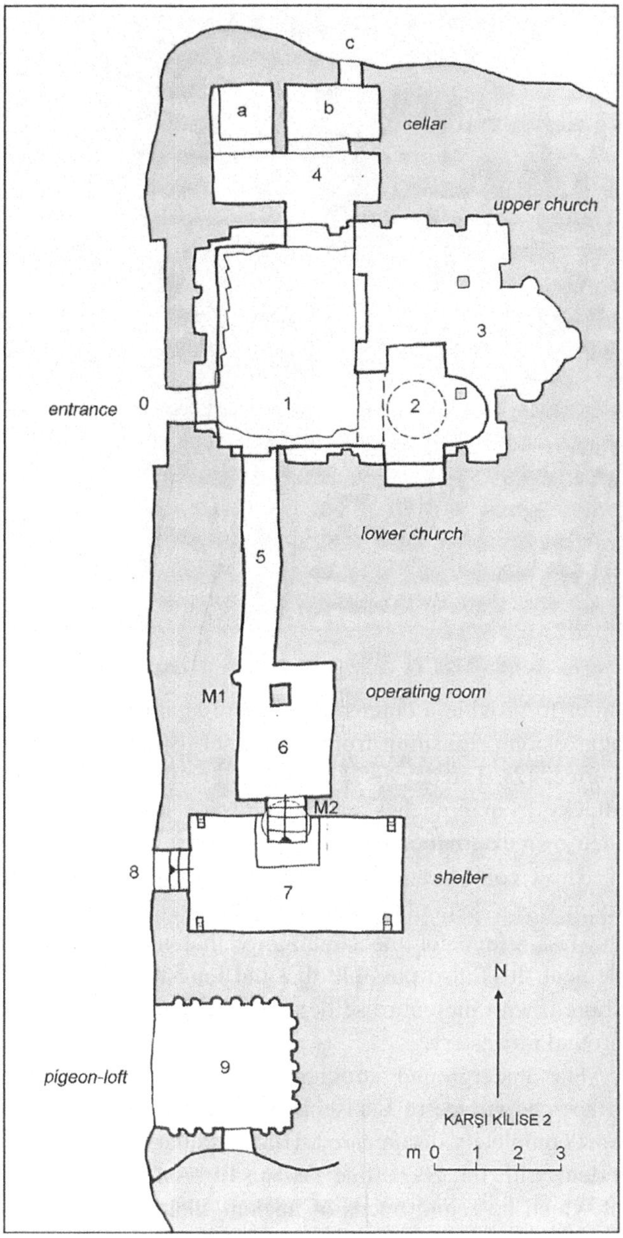

Fig. 4.41 Karşı Kilise Underground Shelter No 2 plan (Bixio, 2012)

passages, which have collectively obscured the original urban design of the shelter. Nevertheless, it is feasible to discern a number of architectural elements that may elucidate the purpose of the subterranean network.

The walls and floors of the rooms and corridors contain niches of varying dimensions, while grain depots are located at the floor level of the rooms or within the walls. Additionally, other types of stores have been identified, with some rooms exhibiting evidence of former use as wine, oil, or salted meat storage. Other rooms with compartments were utilized for the housing of livestock. Additionally, there are wells. These adaptations permit the temporary habitation of a community, thereby ensuring the provision of sustenance.

In the subterranean settlement, a number of millstone doors have been identified, in addition to which the majority of rooms have a groove carved into the rock at their entrance, which would have been used for the placement of a wooden door. The presence of these grooves is particularly prevalent in Kaymaklı (Figs. 4.42 and 4.43).

The settlement's defensive infrastructure is seamlessly integrated into the overall network. This is a vast subterranean complex, designed to safeguard an entire community, their livestock, and their food stores. As with the Derinkuyu complex, this underground structure contains rooms that were used by the community. These include facilities such as underground oil mills, which would not have served a practical purpose if the structure were simply a shelter. It can thus be posited that the subterranean refuge, despite its defensive function, was also utilized during periods of peace as an adjunct to the surface settlement, thereby substantiating the generic designation of "underground store-village" as proposed by Giovannini (1971).

Taking into account the indication of Nicoletti (1980):

> …in Kaymaklı the road-tunnels open into squares surrounded by community rooms, branching in smaller corridors that connect clusters of rooms, perhaps the residences of family groups.On top of the small hill on which the underground settlement was constructed, a necropolis comprising numerous tombs cut into the rock formation is located. The presence of a church, despite the absence of frescoes, within the network indicates that the refugees were Christian. This evidence, when considered alongside the numerous identified rearrangements and the relatively underdeveloped defensive organization, suggests that this underground shelter is among the oldest in the country. It can be placed between the eighth and tenth centuries, a period characterized by frequent raids as a result of Arab attempts to conquer Anatolia.

4.30 Kemer Ören Underground Shelter

The site is situated at an approximate distance of 40 km to the west of Nevşehir, in close proximity to the village of Ovaören. The country, which appears to be almost flat, is characterized by a large number of small funnel-shaped sinkholes from which the site probably takes its name, Kırk Göz (Forty Eyes). One of the pits is accessible via an open entrance. Through the collapsed ceiling, one can descend a cone of debris into a large underground room [1] with a flat ceiling. The original entrance [2] is located at the opposite end, where another crater provides access to a wide descending tunnel.

The corridor on the right terminates at a landslide that exhibits an arch constructed of substantial, regular blocks of stone [3]. A low tunnel crosses a small room at an angle of 90°, allowing for a return to the main hall. This structure is

Fig. 4.42 Kaymaklı Underground Shelter (photo A. E. Keskin)

distinguished by a double row of pillars hewn into the rock. Two tunnels, labeled [5] and [6], are located on the right side of the site. These have been flooded after a few meters. Two additional rooms of modest dimensions are located on the left side [7] and [8]. On the terminus of the corridor, a tunnel, safeguarded by a millstone door that remains in situ, opens into two chambers [9] and [10], from which three tunnels commence, immediately filled with earth.

Approximately 200 m to the northwest of this settlement, there is a necropolis with an underground church (trikonchos with dome) (Hild & Restle, 1981). The principal chamber is reminiscent of a basilica, exhibiting a tripartite division and two rows of substantial pillars hewn into the tufa. In addition, he states that there are two small working rooms with a wine press, which may correspond to rooms [7] and [8].

It seems reasonable to posit that beneath the campaign level, there exists a vast complex of rooms, partially collapsed, which corresponds to the aforementioned sinkholes.

The two tunnels, which were excavated on the right side, appear to have been constructed with the purpose of capturing groundwater for the hydrological needs of the settlement. It can therefore be surmised that the water table would be at a shallow depth from the surface, at approximately five meters. This would suggest that the settlement extends horizontally in the "epidermal belt," without any underlying layers. It is regrettable that the underground tunnels, which originate from this room and are filled with earth, currently prevent further investigation.

The two millstone doors, discovered in a knocked-down state within the hall, were presumably situated within the double entrance of the tunnel [4], at points M1 and M2, in order to provide a defensive measure against any potential aggression originating from the tunnel [3] or from external sources [2]. The device M3 was designed to protect against attacks from branches in the opposite sector, which had their own external entrances (Fig. 4.44).

These considerations lead to the conclusion that the long room was a "redoubt," that is to say, a central area used as the last refuge for the community that inhabited the settlement. It is also possible that the long room served as a church, with the entire settlement functioning as an underground monastery.

The underground structure is located in the subsoil where the village of Eskişehir once stood. This village has now completely disappeared. Hild considers this to be coincident with the Byzantine Osiana (or Asiana, or Osiena), of which he found traces of ancient shelter walls (Hild & Restle, 1981).

4.31 Kırlangıç Valley Underground Shelter

Kırlangıç Valley Underground Shelter is located approximately 1 km north of the Hisarcık District in Kayseri. It is positioned along the western wall of the canyon-like Kırlangıç Valley, which stretches from Hisarcık all the way to Mount Erciyes. The strategic location of this underground structure, overlooking the valley, suggests that it may have played a significant role in the region's history, whether as a refuge, a settlement, or a structure with a unique purpose (Fig. 4.45).

This underground shelter is an impressive feat of engineering, extending to a total length of 505 m with a depth

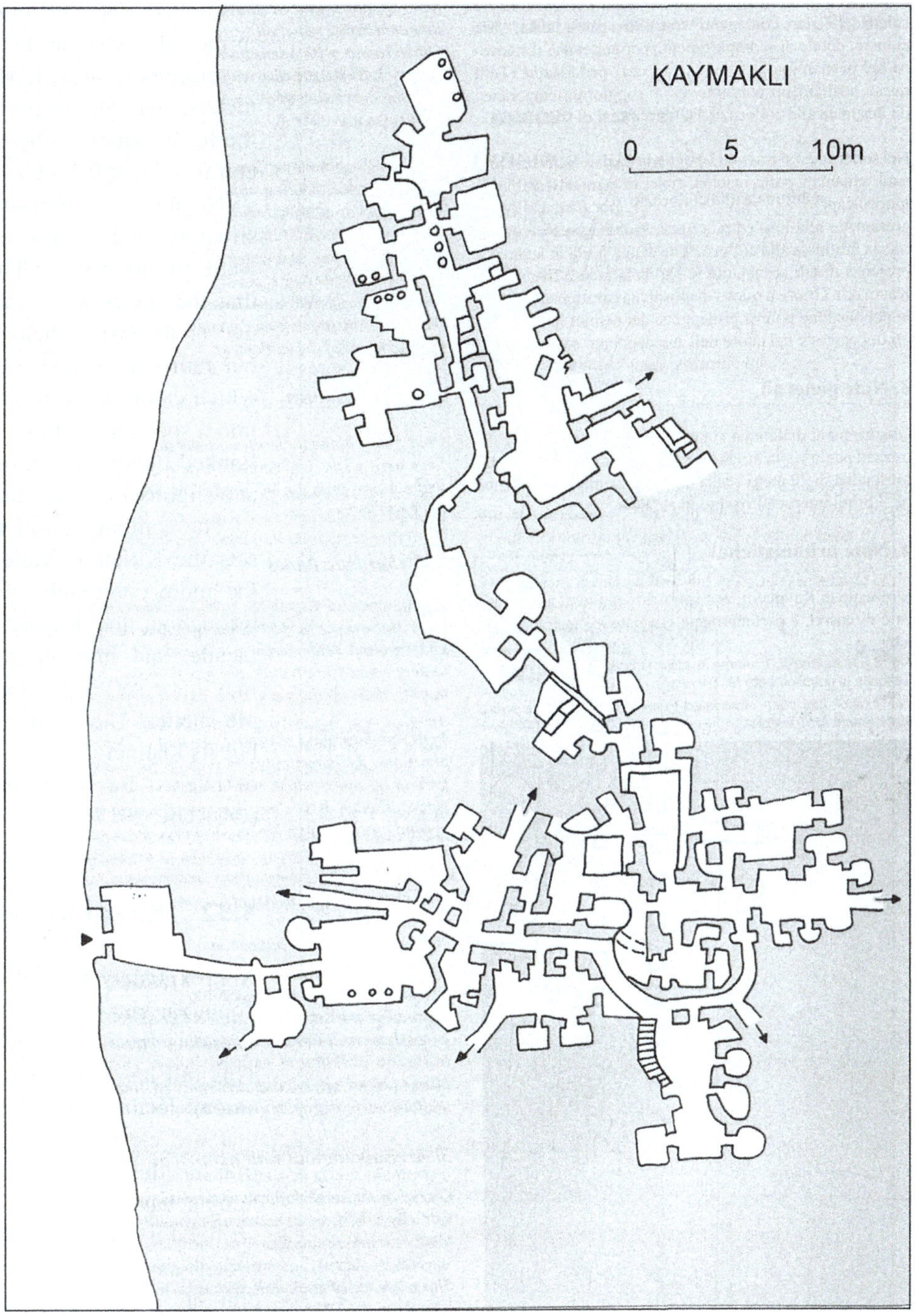

Fig. 4.43 Kaymaklı Underground Shelter plan (Bixio, 2012)

of 56 m. What sets it apart from other similar structures in the region is its distinctive architectural design—unlike many underground complexes that expand horizontally, this one has been carved vertically across five different levels. The most remarkable aspect of its construction is that the entire front façade remains open to the valley, providing natural light and ventilation, a rarity among such rock-cut dwellings.

The first three stories of the structure, accessible via a sloping passage, primarily consist of large halls that contain multiple living rooms. These levels seem to have been designed for habitation, providing sufficient space for daily life and communal activities. The fourth floor, however, features smaller rooms that are accessible via steep stairs, indicating a shift in function from the lower levels. The final floor consists of a blocked tunnel, which is believed to extend downward to the valley floor, although its precise endpoint remains uncertain.

Further investigations within the valley revealed the existence of another tunnel entrance carved into the rock,

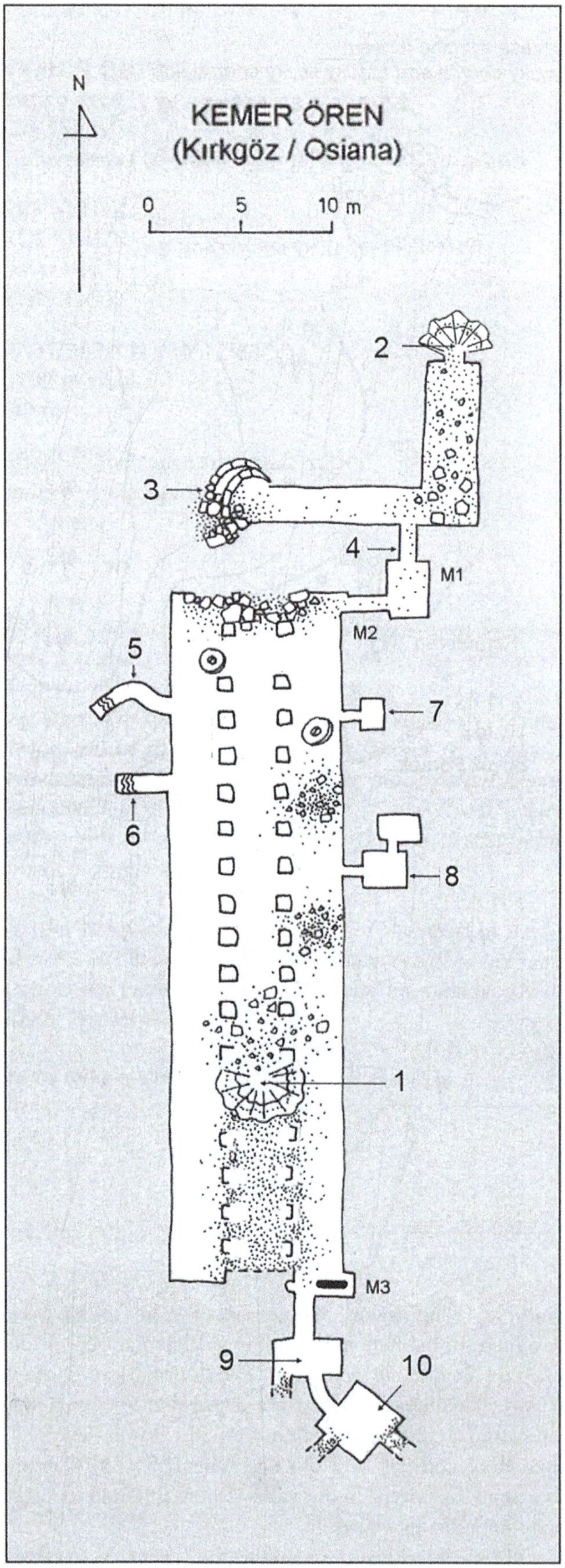

Fig. 4.44 Kemer Ören Underground Shelter plan (Bixio, 2012)

positioned at the level of the creek. This entrance is located in the same general area as the underground shelter. Given its placement at a depth of approximately seven to eight meters, it is highly probable that the blocked tunnel from the lowest floor of the structure originally extended to this entrance, establishing a connection between the two points.

Notably, no defensive gate has been identified in the structure, and each level contains windows that open outward to the valley. These architectural features suggest that the site was not merely designed as a temporary refuge or defensive shelter, as many similar structures were, but rather as a more permanent settlement. The entrance, which can only be reached via a highly steep slope, opens into a spacious hall of considerable size. This central hall is flanked by six meticulously carved rooms, each positioned at a slightly lower elevation.

The combination of these unique architectural elements sets the Kırlangıç Valley Underground Shelter apart from the more commonly observed rock-cut settlements found throughout the Kayseri region. Its vertical design, open façade, and intricate room arrangements indicate that it may have served a special function beyond that of a simple shelter. The precise history of its use remains unclear, but its distinct characteristics continue to make it a subject of interest for researchers and historians alike (Fig. 4.46) (OBRUK, 2017).

4.32 Kirkor Underground Shelter

Approximately five kilometers to the north of the town of Tomarza, situated just to the west of the main road that connects the area, stands a distinctive and prominent hill. This hill has long been a subject of curiosity and interest among locals, researchers, and visitors. Unlike most hills that are clearly natural formations, this particular elevation has the peculiar appearance of having been shaped by human hands. It gives the strong impression of being an artificial mound, rather than a product of natural geological processes. The local community has referred to it for generations as the "Kirkor Place," a name that has become synonymous with the area's mystery and historical richness.

Over time, this location has yielded a considerable number of archaeological finds. Chief among these discoveries are fragments of pottery scattered broadly across the surface of the hill and its surrounding plains. These remnants strongly indicate that the site was once inhabited or actively used over an extended period. The consistent presence of these artifacts points to a continuous or repeated pattern of human activity, possibly extending back several centuries, or even millennia. As such, the area holds particular significance not only for local historians but also for

Fig. 4.45 Entrance of Kırlangıç Valley Underground Shelter (photo A. E. Keskin)

archaeologists and heritage enthusiasts who recognize its potential to provide insights into the region's past.

Beyond its surface-level interest, the site is also home to a number of buried historical features. Located on the open plain just south of the hill are several ancient graves, believed to be burial sites dating back to a distant past. More notably, however, is the presence of an underground shelter of remarkable construction. Though positioned relatively close to the main road, the entrance to this underground complex is exceptionally well hidden. It is cleverly camouflaged by the natural terrain, making it difficult to identify even for those actively searching.

Once the discreet entry is located and accessed, visitors step into an initial chamber with an area of 65 m^2. This main room once had an impressive security mechanism in place, as evidenced by the presence of a large millstone door that once served to block the entrance and protect those inside.

From this primary chamber, a network of tunnels extends in multiple directions, forming a complex and well-planned subterranean structure. To the north, a passage runs for 38 m before terminating at a blocked flue, which now prevents further entry. Heading south from the central chamber, another corridor stretches for about 10 m, leading into a significantly larger room. This large chamber is connected to six smaller side rooms, likely used for a variety of domestic or defensive purposes.

Continuing even further southward, a narrow tunnel travels for an additional 68 m before branching into a T-shaped junction. The western arm of this junction runs another 34 m but narrows considerably, making it impassable beyond a certain point. In contrast, the eastern branch leads into a set of three spacious, interconnected chambers. Among them lies an especially intriguing feature—a deep well carved into the rock, plunging 14 m underground, likely used as a vital water source for the shelter's occupants.

The complex and carefully constructed nature of this underground settlement points to its historical use as a secure refuge. It stands as a testament to the ingenuity and resilience of those who built it, reflecting the defensive strategies of communities long past (Figs. 4.47 and 4.48) (Yamaç, 2018b).

4.33 Kömür Underground Shelter

The Kömür Underground Shelter is located in a narrow, rocky valley approximately 2 km south of Kömür Village, which belongs to the Tomarza District. This underground structure, hidden within the natural contours of the valley, stands as a notable example of historical defensive architecture. Carved meticulously into the earth, the shelter showcases a complex layout that hints at its significance in the past. The design and features of the shelter suggest that it was not merely a hiding place but served a more structured and strategic function, possibly as a safe haven during times of danger or invasion.

One of the most distinctive and visually striking features of the structure is the ceiling of the main hall, which includes a vertical shaft. This pit, measuring 5.70 m in

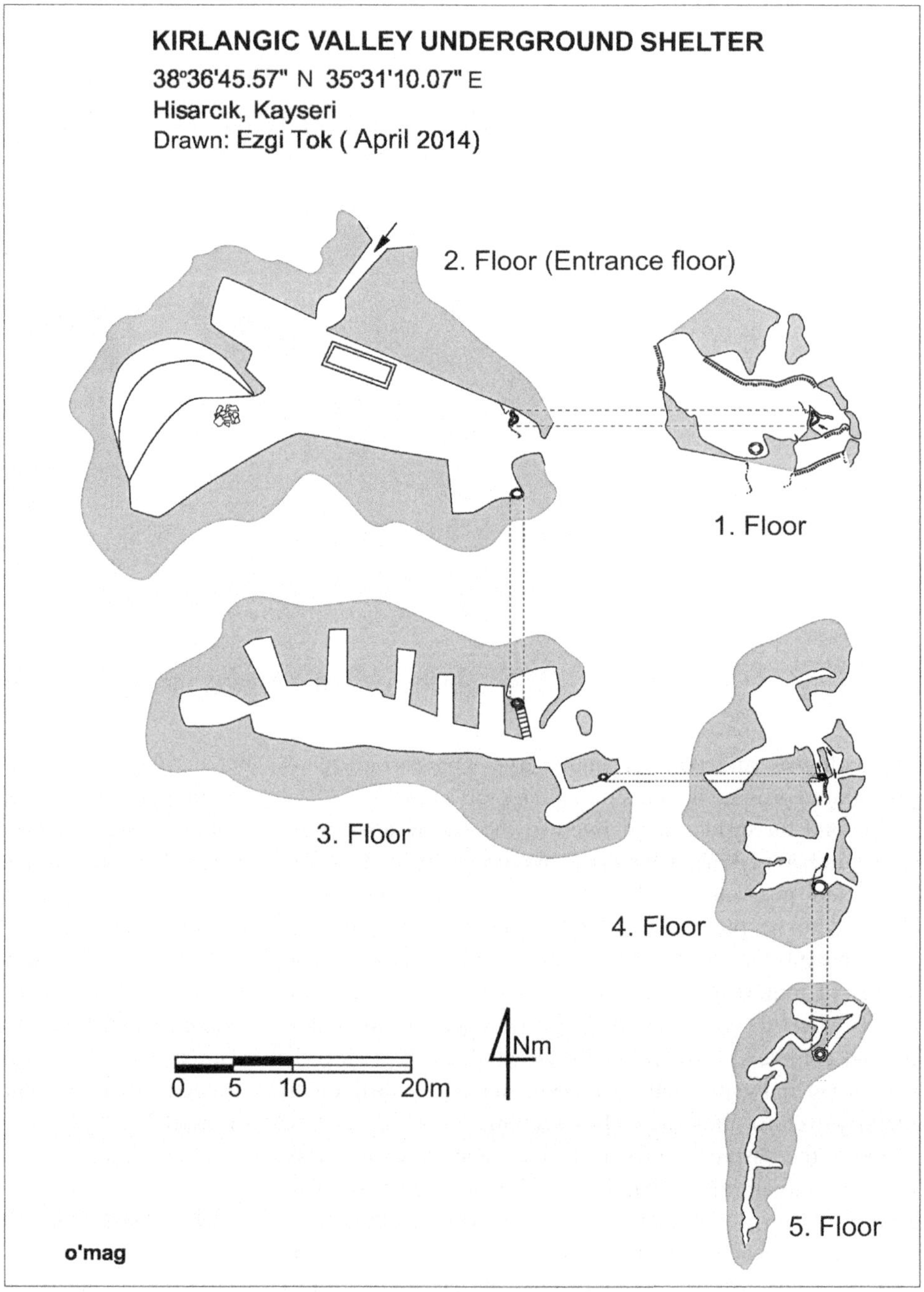

Fig. 4.46 Kırlangıç Valley Underground Shelter plan (drawn E. Tok)

height and 1.70 m in diameter, rises directly upward from the chamber below. Carved into the walls of the shaft are a series of foot and hand steps, indicating that it was once used to climb between different levels of the shelter. Such features reflect the thought and craftsmanship put into the construction, allowing for vertical movement within the structure in addition to the horizontal tunnels and rooms.

The entrance tunnel that provides access to this central chamber is protected by a large millstone door, a hallmark of defensive underground settlements. These heavy stone doors were crucial in securing the entrances, acting as barriers against unwanted intrusions. Interestingly, the chamber's ceiling also features an unusual opening that leads directly to the surface. Although it appears to serve no immediate or functional purpose, this pit may have been created later, perhaps during a repurposing of the shelter after its original use had ended. Its irregular design and placement suggest that it was not part of the initial architectural plan. Due to its structure, this main hall would be extremely impractical to use as a storage space or silo in its present state, further supporting the idea that modifications may have occurred in later periods.

Fig. 4.47 Kirkor Underground Shelter (photo A. E. Keskin)

The Kömür Underground City also includes a blocked tunnel that once served as the main entrance to a now partially exposed first chamber. This section has suffered from collapse, which has revealed parts of the room beneath. Like the entrance to the main hall, this tunnel was also protected by a millstone door, reinforcing the consistent emphasis on defense throughout the complex. From the main hall, a narrow tunnel extends northwest for about 32 m, rising in elevation from its starting point. This passage, also defended by a millstone door at its beginning, ultimately leads to three rooms. However, these rooms have been heavily damaged and are now exposed to the surface due to collapse.

In addition, there is another tunnel that branches off to the east of the main hall. Unfortunately, this passage is blocked after a short distance, preventing access to any further areas that may lie beyond.

Despite the extensive damage and the passage of time, the Kömür Underground Shelter remains a compelling example of ancient engineering and strategic design. Its multi-layered layout, defensive features, and architectural complexity continue to provide valuable insights into the lives and practices of those who once sought refuge within its walls (Figs. 4.49 and 4.50) (Yamaç, 2018b).

4.34 Kuşcağız Underground Shelter

Kuşcağız Underground Shelter is located in a quiet, narrow valley approximately 3 km east of Kuşcağız village in the Tomarza district of Kayseri Province. This structure stands out as one of the most substantial and well-constructed underground defensive shelters in the entire region. Its strategic position within the valley and the care taken in its construction demonstrate its historical significance and suggest that it once played a major role in local defense systems.

The shelter features three separate entrances that are located along the slope facing the valley. These entrances are partially hidden and surrounded by rocks, many of which have collapsed over time, making them more difficult to identify. A fourth entrance is located further down, positioned directly in front of the valley. Each of these four entrances is protected by millstone doors that remain in place to this day, highlighting the advanced defensive planning involved in the shelter's original design.

Although it is evident that a settlement of rock dwellings once existed in and around the valley, the full size and layout of the settlement are still unclear. Much of the original structure has been damaged or lost due to the collapse of the surrounding rocky walls, which limits the current ability to fully explore the area. Despite this, traces of habitation are still visible, and it is clear that this area once supported a substantial community.

What makes this site even more fascinating is the presence of numerous rock tombs and burial sites carved into the valley walls. At least some of these are believed to date back to the Roman and Byzantine periods, offering strong evidence that the valley has been continuously occupied or used for centuries. It is especially interesting that the Kuşcağız underground shelter, like the nearby Kirkor and Kömür underground shelters, was carved into a distinctly funerary zone. This suggests a longstanding and possibly spiritual or ritual connection between the defensive structures and the burial areas within this region.

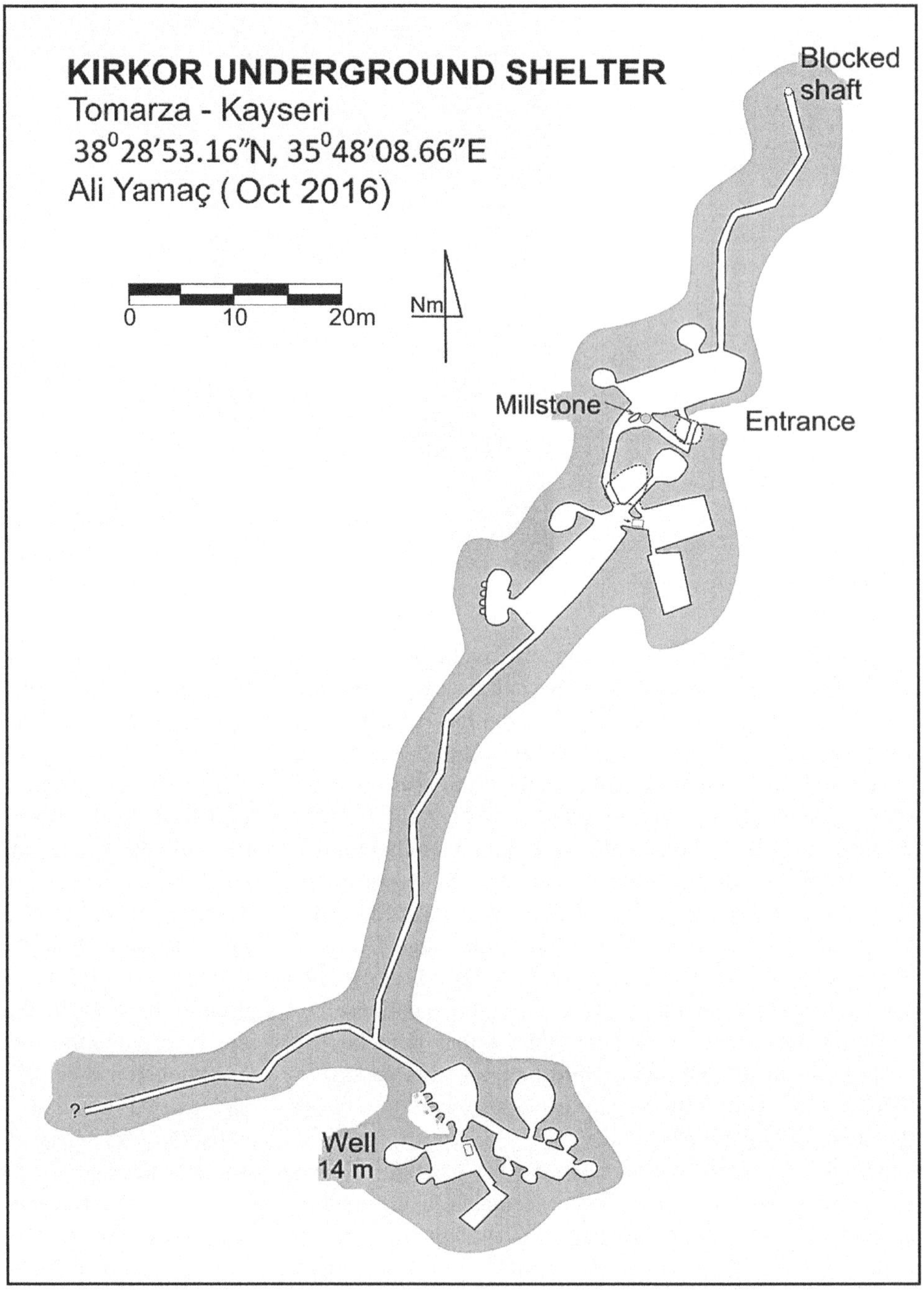

Fig. 4.48 Kirkor Underground Shelter plan (drawn A. Yamaç)

The Kuşcağız Underground Shelter stretches approximately 105 m along its north–south axis. Inside the shelter, there are a total of ten chambers, several of which are particularly remarkable in terms of size and design. One of the most notable rooms is located just past the second entrance on the northern side. This chamber, which lies at a slightly lower level than the entrance tunnel, measures an impressive 13 by 7 m and likely served a significant function. Another chamber, located in the northern part of the shelter and believed to be used for storage, contains eleven underground silos—an indication of the careful planning involved in the shelter's construction.

While excavation appears to have stopped in the southern tunnels of the shelter, two of the tunnels extending to the north are currently blocked with debris. It is quite likely that the underground complex extends further in that direction, and additional chambers or features may still be hidden beneath the surface.

The Roman and Byzantine ruins visible on the surface further reinforce the historical importance of the valley.

Fig. 4.49 Kömür Underground Shelter (photo A. E. Keskin)

Given its rich archaeological potential, the entire area is in need of a much more detailed and comprehensive study (Figs. 4.51 and 4.52).

4.35 Mazı Underground Shelter

Mazıköy (formerly Madazi) is identified with the ancient Mataza (Hild & Restle, 1981). The town is nestled along the bottom of a canyon, crossed in the center by the road from Kaymaklı. On the high portion on both cliffs' sides, at the entrance of the village, there are five rocky tombs. Of these, one has an entrance with a double frame. Another four are of the "portico" type with Doric columns, dating from the period of the kings of Cappadocia (334–30 BC). The pre-Hellenistic settlement of Mazıköy is evidenced by steps and platforms carved on top of the plateau. Here there are also many graves dug in the floor of the rock. Instead, the rocky tombs with monumental character, dated to the Roman period, are located in the eastern part of the canyon.

The underground shelter is located in the center of the village. It was researched and mapped by the members of the National Commission for Artificial Cavities of the Italian Speleological Society in 1993 and by OBRUK Cave Research Group in 2013.

This underground shelter is open for visitors and it is believed that it extends under all of the houses in Mazı Village. There is another underground shelter as large as Mazı Underground Shelter under the houses in the north of the section that could be entered. These two different systems could be linked to each other. However, it is not possible to see the connection due to the collapsed, blocked passages and the walls between them.

The area accessible to tourists has two distinct levels. Starting from the entrance, the rooms stretching to the left and right were utilized throughout the tranquil period.

The accessible section consists of two overlapping tiers, each with distinct entrances. The initial floor system features a singular entrance and comprises a sequence of six rooms excavated consecutively along a uniform axis, with the exception of a lateral compartment, linked by a brief corridor. The final chamber is preceded by a millstone door. Two tunnels obstructed by earth originate from halls.

The upper floor has greater complexity. From the entrance, we can go via a sequence of rooms aligned along an axis oriented toward the southwest, leading to a junction. A narrow passage on the right leads to a partially collapsed chamber with two rooms and a junction of tunnels. To the left, the tunnel extends orthogonally.

On either side of this axis, facing southeast, there are small chambers and connecting tunnels that create an open network. One of the corridors extends southwest to a wider area from where an uncharted tunnel originates. The path proceeds from the hall to the chamber, incorporating two perpendicular directional shifts. The final chamber is preceded by a millstone door (Fig. 4.53).

Additional holes have been excavated on the cliff, rendering them currently inaccessible. The various levels were linked by thin internal vertical shafts featuring footholds. The chamber in the final in the sequence, secured by the millstone door, served as the "redoubt," denoting the more fortified subterranean sanctuary.

The top system requires further specific investigations, not just to continue tunnel exploration but also to identify further closing devices beyond those found at point. It is quite probable that the subterranean construction contained a minimum of three additional millstone doors to facilitate the isolation of the core region from any direction in the event of a threat (Bixio, 2012; Yamaç & Tok, 2015c).

4.36 Meskendir Underground Shelter

The Meskendir Underground Shelter, located in the Meskendir Valley near Göreme, Cappadocia, is a multi-tiered rock-cut complex that evolved over time, adapting to various defensive and domestic needs. The site showcases different phases of construction, repurposing, and

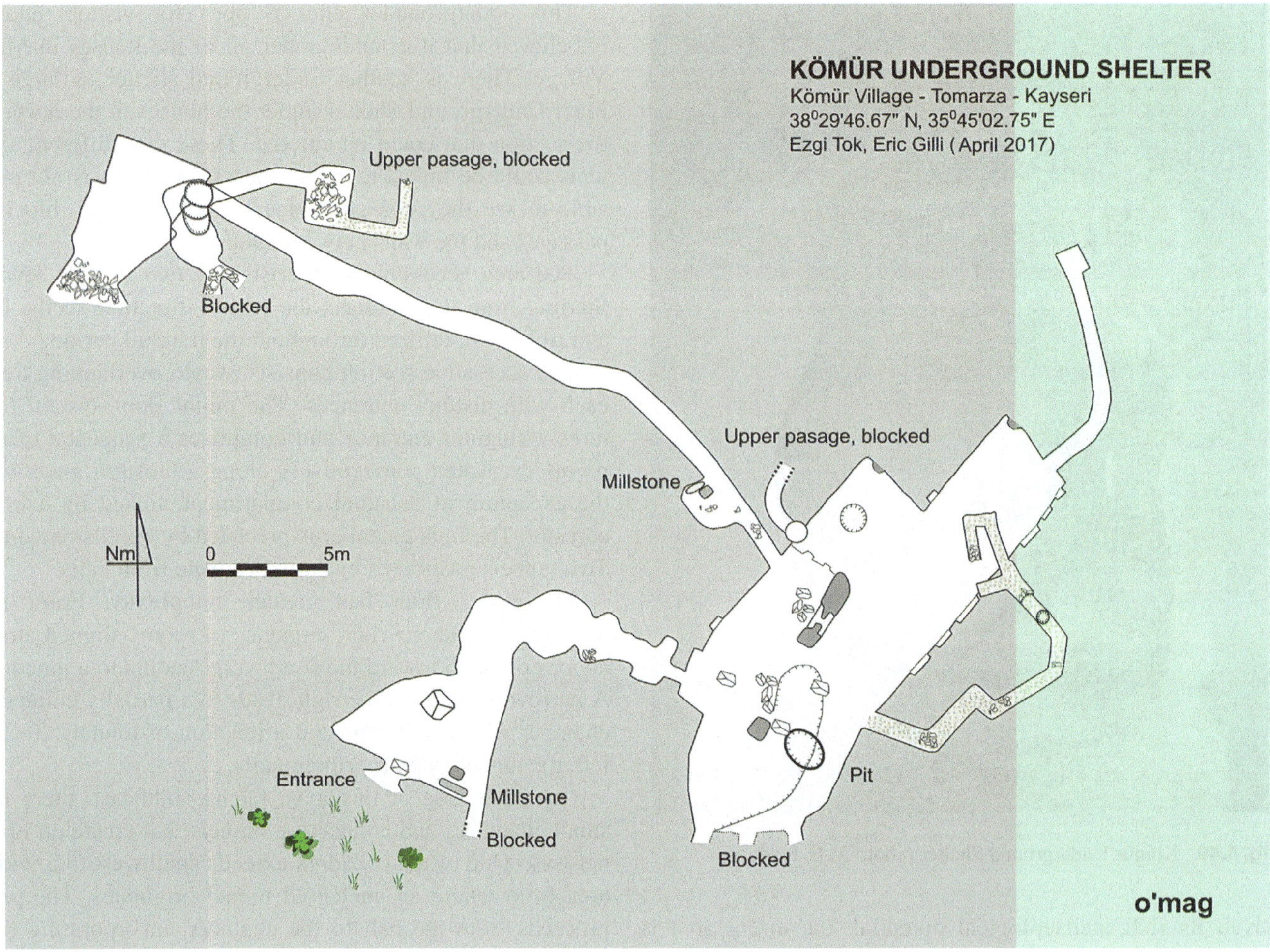

Fig. 4.50 Kömür Underground Shelter plan (drawn E. Tok, E. Gilli)

functional transformation, reflecting the broader historical development of Cappadocian underground shelters.

The oldest part of the shelter, Room 1, was a single chamber accessible via a narrow tunnel secured by a millstone-shaped door. This design was a typical defensive feature, allowing inhabitants to retreat and barricade themselves inside. The compact space suggests that it was originally a small refuge rather than a full-fledged underground settlement.

Over time, the shelter expanded significantly, with the addition of Rooms 2–9 forming the second tier. This phase introduced more advanced defensive mechanisms, including inclined passages, vertical shafts, and additional millstone doors. These improvements suggest an increased need for protection, likely due to external threats.

The presence of vertical wells and narrow corridors, which would have been difficult for attackers to navigate, indicates that the shelter was designed for prolonged habitation under siege. The defensive complexity implies that this phase was a response to a period of heightened instability.

The third and fourth tiers introduced new features such as embrasure-like windows and a rock-cut toilet. These additions hint at an evolving defensive strategy, possibly incorporating active defense elements like archers using the openings for visibility and attack.

However, there is little evidence of prolonged habitation in these upper levels. They may have served as temporary refuges or emergency escape routes rather than permanent living quarters.

In a later phase, the defensive elements were dismantled, making the shelter more accessible. Millstone doors were removed, and corridors widened, transforming the site into a more practical space for daily activities rather than a refuge. This period saw the repurposing of lower-tier rooms for domestic and utility functions, such as:

- Animal stalls: Indicating agricultural use.
- Fireplaces: Suggesting food preparation and heating.
- Storage areas: Likely used for grain and supplies.

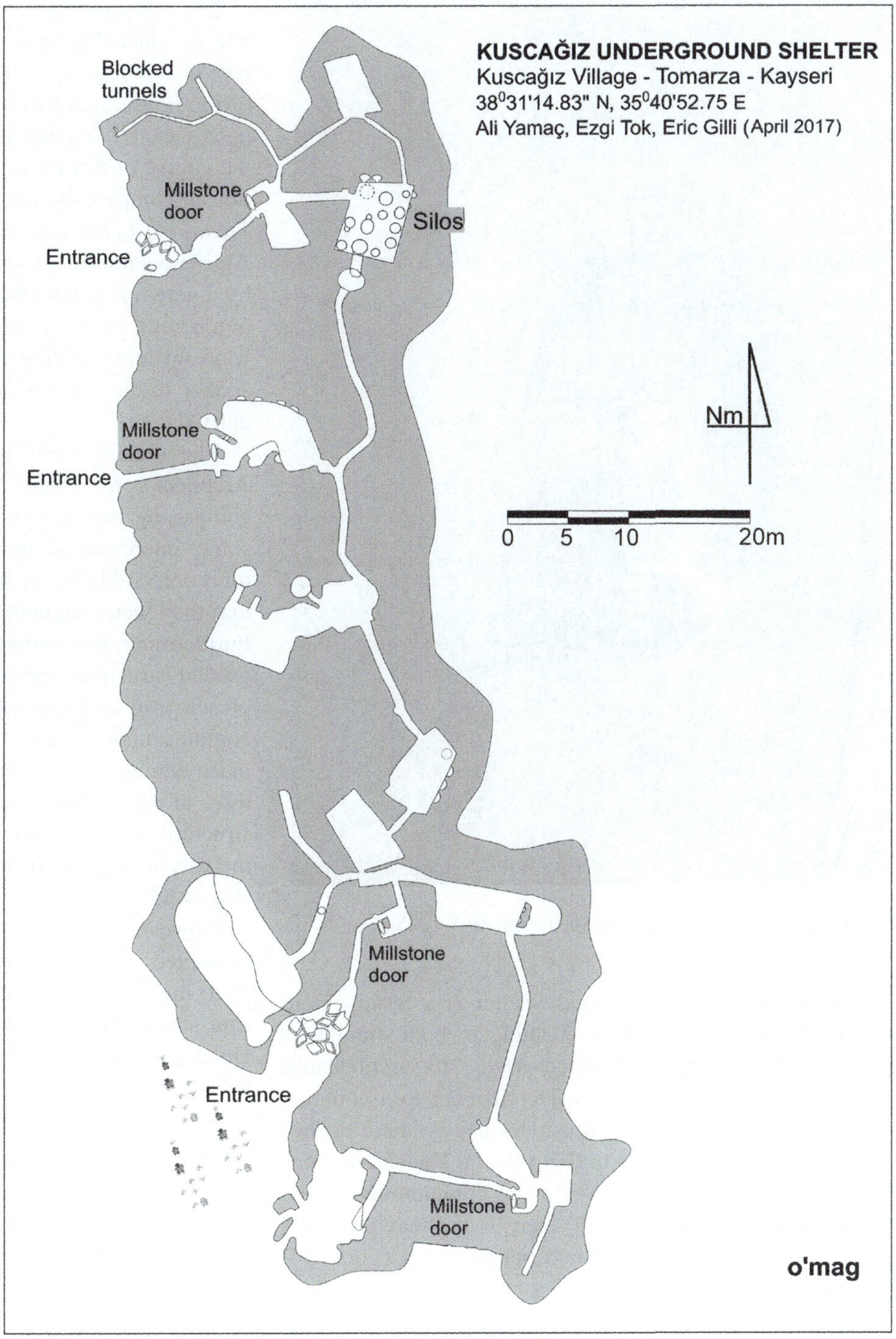

Fig. 4.51 Kuşcağız Underground Shelter plan (drawn A. Yamaç, E. Tok, E. Gilli)

A drainage system was introduced, improving sanitation. The addition of a small shelter (Structure 11) and the conversion of a chamber into a dovecote suggest continued use even after the site was no longer needed for defense.

The Meskendir Underground Shelter exemplifies the adaptive nature of Cappadocian rock-cut architecture. The gradual transformation from a simple hideout to a complex multifunctional settlement reflects changing historical circumstances, including shifts in security, settlement patterns, and economic activities.

While the exact dating of these modifications remains uncertain, the site's evolving structure aligns with broader regional trends observed in Cappadocia's underground settlements. The transition from purely defensive shelters to more integrated living spaces mirrors the region's fluctuating security concerns over centuries.

Fig. 4.52 Kuşcağız Underground Shelter (photo A. E. Keskin)

The Meskendir Underground Shelter is a testament to the ingenuity of its builders, adapting to both immediate threats and long-term habitation needs. Its architectural evolution—from a single defensive chamber to a complex domestic and utility space—illustrates the dynamic history of underground settlements in Cappadocia. The site remains an important example of how human communities in the region continuously modified their living environments to meet changing challenges and opportunities (Fig. 4.54) (Bobrovskyy & Grek, 2015).

4.37 Mimar Sinan Underground Shelter

Mimar Sinan (Sinan the Architect) is regarded as the most significant architect of the Ottoman Empire, with an estimated 400 structures of historical and cultural value attributed to him. This remarkable architect, who lived and worked under the reign of four Ottoman sultans, was born in 1489 in the small village of Ağırnas, located in the Kayseri Province. The house in which he was born has been restored and is now a museum.

The restoration and environmental planning of this house was conducted between 2006 and 2008 (Sözen, 2004). As part of the ongoing restoration and environmental planning of the house, a number of the underground structures have been cleaned, organized, and opened to the public. During the course of this project, it became evident that the subterranean complex did not merely encompass the residence of Sinan, but rather extended to cover the entire settlement of Ağırnas, with the various dwellings therein interconnected by a network of tunnels. It is regrettable that a considerable number of property owners have obstructed the passageways by constructing masonry walls, thereby preventing access to their residences from a subterranean point of entry.

The underground complex situated beneath Sinan the Architect's residence encompasses an area of 1850 m^2, comprising two stories. As a result of the cancellation of some interconnections, four distinct subterranean structures are visible below Sinan's house. However, it is evident that they were originally part of a single, unified structure. Furthermore, the continuation of the first and fourth underground structures below the neighboring houses and the presence of different exits to the surface indicate that the complex underground structure in Ağırnas Village was in place centuries ago. The presence of four millstone doors, three of which remain in situ, within the fourth underground structure provides clear evidence that the structure was, at the very least, constructed with protective purposes in mind (Fig. 4.55).

However, as the defensive concerns that initially prompted their construction were subsequently alleviated, these underground structures have undergone significant structural alterations. Consequently, the Mimar Sinan Underground Shelter Complex comprises four distinct subterranean structures.

The initial subterranean structure, situated in close proximity to the museum, can be accessed via a substantial chamber whose ceiling is supported by stone vaults. The barrel vault roof, a common structural feature observed in numerous subterranean chambers throughout the edifice, is also present in the subterranean spaces beneath numerous dwellings in Ağırnas Village. As documented by Sözen (1988), this architectural style is likely the oldest structural design in the village. The initial subterranean settlement extends eastward, encompassing a series of rooms and extending beneath the adjacent residential structures, as illustrated in the plan.

The second subterranean structure is located within the primary edifice and can be accessed via a staircase leading from the ground level. The structure is that of a small storage area, with both connections to other neighboring underground structures sealed with stone masonry. The presence

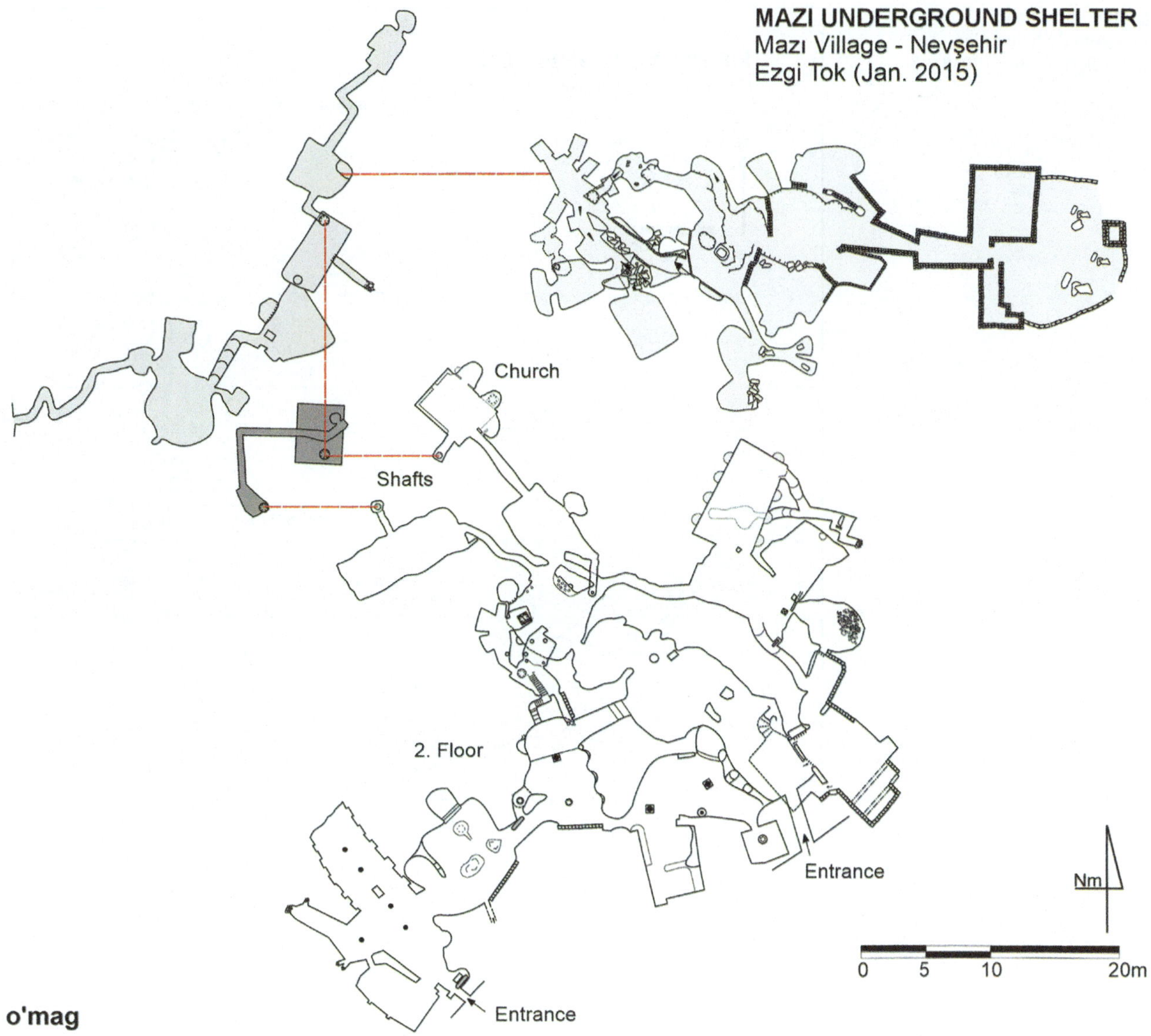

Fig. 4.53 Mazı Underground Shelter (drawn E. Tok)

of holes on the ground in the main room of this small structure, which were used for the placement of jars, indicates that the primary function of this building was storage.

A third underground structure, accessible via a narrow alleyway, is located on the side of the main building and serves as an additional independent storage area. Although it is conceivable that this edifice may have served as a residential space, connected to other subterranean structures, it is evident that this function was abandoned centuries ago, and it subsequently became utilized for the storage of items.

The fourth underground structure, the largest and most significant of all those beneath Sinan's residence, merits particular attention. The entrance to this subterranean structure is located at the end of the aforementioned alleyway. In a chamber with a barrel vault roof located toward the south of the aforementioned entrance, furnaces are present, which are believed to be an iron foundry. Subsequently, the subterranean structure extends eastward, emerging from the main thoroughfare after traversing an internal residence. Its initial configuration encompasses two levels, and it has been refurbished and opened to the public.

Subsequently, the route proceeds underground, beneath other residential properties, via a narrow passageway. The subterranean sections have not yet been cleaned, and, more intriguingly, some are still utilized by the occupants of neighboring dwellings. The final gallery, which extends eastward, is obstructed by detritus, and another millstone door serves to reinforce the initial construction objective of the structure (Fig. 4.56) (Yamaç, 2023; Yamaç & Tok, 2015a).

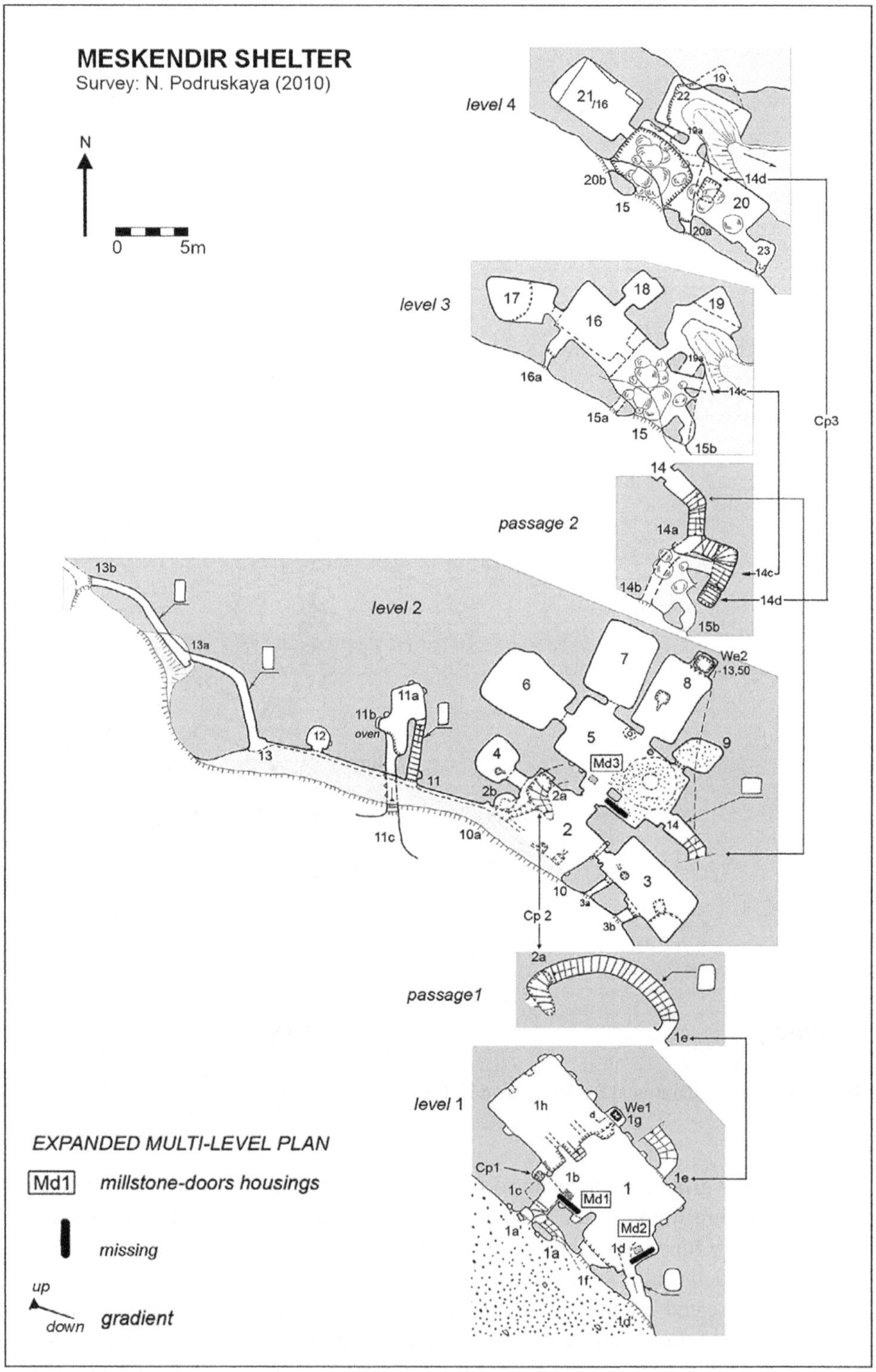

Fig. 4.54 Meskendir Underground Shelter plan (Bobrovskyy & Grek, 2015)

4.38 Mucur Underground Shelter

The underground shelter is located in Solaklı Quarter of Mucur District, 21 km north of Kırşehir Province. The underground site is located among the modern houses that make up the quarter. Several entrances located on the ground level allow access to the two floors of the underground structure currently known, which are open for touristic visits.

It is believed that the underground shelter goes deeper and covers a larger area than its current state. There are up to eight levels, reachable following further works of

Fig. 4.55 Mimar Sinan Underground Shelter (photo Ç. Çankırılı)

excavation deep down. The first floor consists of 14 rooms of various shapes ministering to different purposes. All the rooms are linked to each other by a network of tunnels that looks like an underground road system. On the first floor, there are 12 millstone doors, of which seven are in situ.

The irregular planned second floor extends in the east–west direction and consists of 8 different rooms. The most interesting room on this floor is Room P. The room is rounded and all the walls are covered with niches. Although this room looks like a dovecote, it has no connection from inside to outside. This floor contains two millstone doors as well.

The underground settlement is defended by many closing devices. In total there are, in the accessible areas, no less than twelve operation rooms on the first floor, seven of which are still equipped with their millstone doors, while the other five have been removed. In the second floor at least four operation rooms with two millstone doors on site have been identified.

In comparison to the numerous underground shelters that were investigated in the neighboring region of Gülşehir, situated to the south of the Kızılırmak river, this settlement displays a more complex architectural layout, which can be categorized as an example of an underground city. Therefore, it can be concluded that this is the northernmost underground city to have been identified within the territory of Cappadocia. This assertion can be further substantiated should the existence of additional floors in the deeper levels be confirmed.

About the closing devices, it should be noted that some of them have been designed to allow the defense in both directions of the entrances of the tunnels. This presupposes the existence of underground systems connected, but with independent entrances from outside.

It should be noted, finally, the circular chamber (P), with overlapping niches, similar to a dovecote, located on the second level, is the only one of this form (bell-shaped) identified so far in Cappadocia (Fig. 4.57) (Çakır et al., 2016).

4.39 Örentepe Underground Shelter

The road from Acigöl to the south extends for 15 km to reach Kurugöl Village. The journey continues southward along the compacted road that marks the western edge of a range of hills. Upon covering approximately 3 km, one encounters a prominent hill, distinguished by its rounded contours. This site is designated as Örentepe. The western bare slopes of the hill were traversed on foot. At the midpoint of the ascent, a series of artificial excavations are observed, positioned approximately along two intersecting lines. These openings grant entry to a significant underground edifice.

It is essential to evaluate whether this settlement aligns with the one described by Martin Urban. The site is located approximately 2 km northwest of the village of Kuyulutatlar, as indicated on the Urban map, along the route connecting Derinkuyu and Güzelyurt. This location is analogous in nearness to the Örentepe site. Although Urban's Örentepe Underground Shelter plan (Fig. 3.6) (Urban, 1969) and the plan drawn by Bixio (Fig. 4.58) do not show much similarity, these two underground shelters must be the same place.

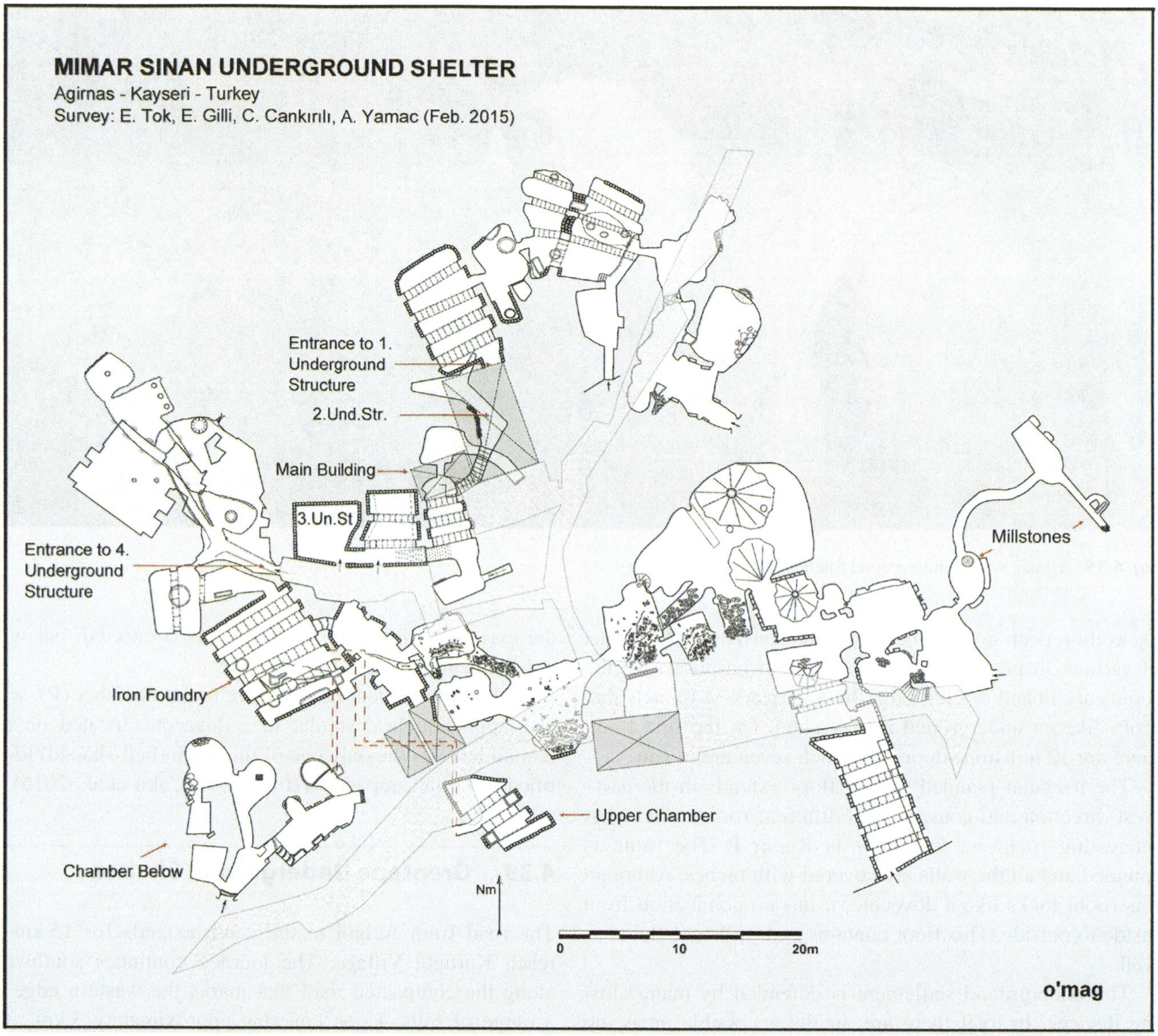

Fig. 4.56 Mimar Sinan Underground Shelter (drawn E. Tok, E. Gilli, Ç. Çankırılı, A. Yamaç)

Several entrances are positioned at two different elevations. The vertical elevation difference is 5 m. They appear as a substantial crater filled with soil rather than constructed edifices. Several entrances still connect to the subterranean network that resembles an authentic labyrinth. Each aperture in the slope connects to the chambers hewn from the rock. Each of them extends to the corridors linked to the series of rooms that stretch from the excavated area to the interior sections. Certain rooms feature elevated corridors connected to the network on the upper floor, which exhibits a comparable layout to the lower floor. Consequently, all the subterranean networks resemble brief corridors linking them together.

Some of them still permit access to the subterranean network, which is unveiled as an intricate labyrinth. Every aperture that leads to the slope is linked to a chamber hewn from the rock. Each of them leads to additional corridors, occasionally on both sides, which connect to the subsequent chambers excavated in deeper regions. Consequently, certain rooms are fitted with ascending passages that connect to the net situated at the upper level, which mirrors the layout of the lower floor. Consequently, all subterranean regions seem to be interconnected by brief passageways. No tunnels exceeding 5/6 m seem to have been developed. The defense structures appear rather unassuming.

The plan presented here, derived from a sketch, constitutes merely a minuscule portion of the examined network. Nonetheless, it serves as an exemplary illustration of the scheme that informs the settlement. The correlation between the voids (i.e., the excavated chambers) and the

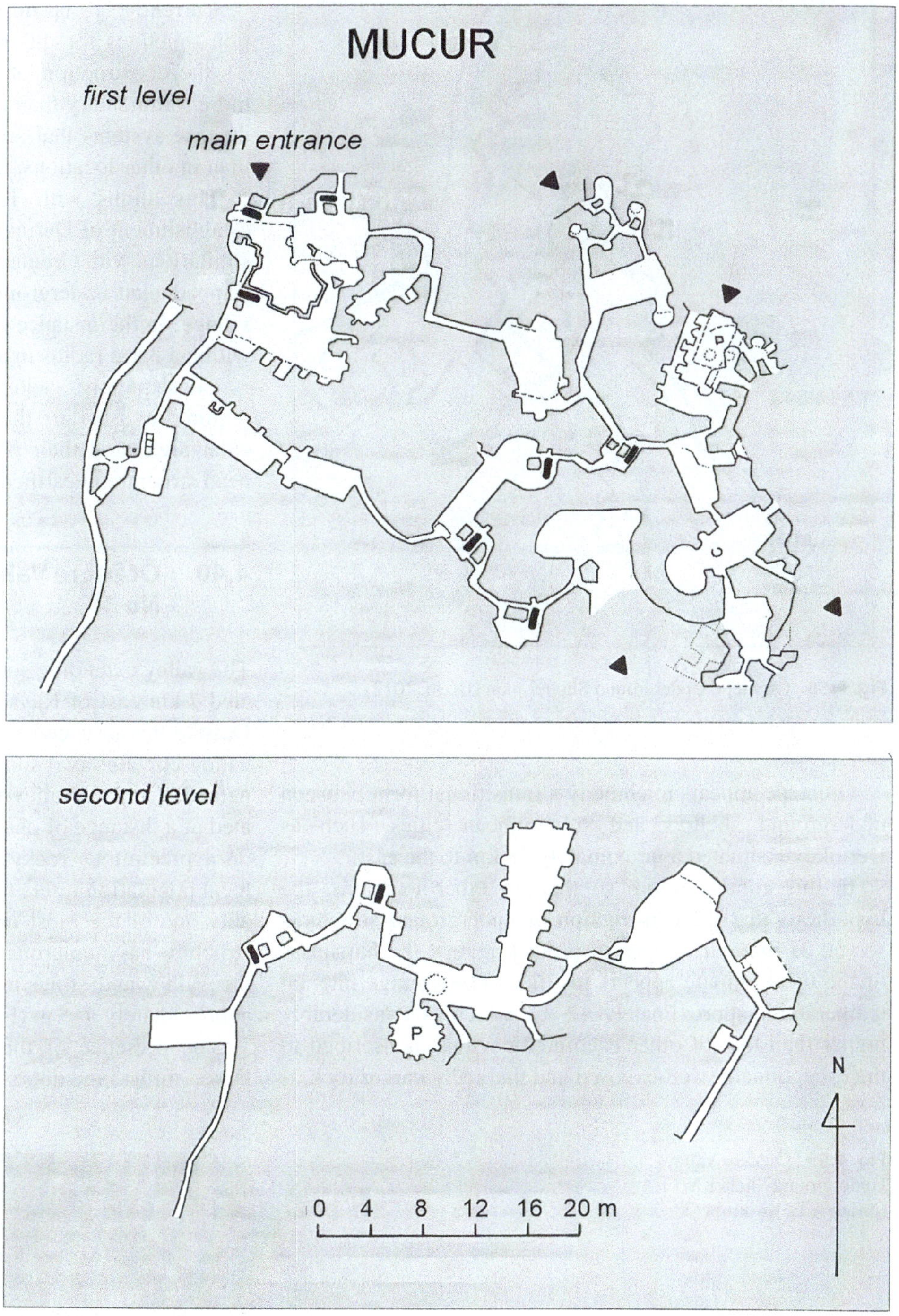

Fig. 4.57 Mucur Underground Shelter (Bixio, 2012)

rock mass is significantly more pronounced than in any other subterranean shelter in Cappadocia, characterized by a configuration of the chambers that appears less hierarchical than in other locations.

The absence of extensive tunnels, which are characteristic of other structures and serve as escape routes or connections between satellite systems for enhanced mutual defense, is noteworthy in this exploration. Millstone doors appear more restrained and constrained. The floor plan indicates that there is a singular operational chamber (M1) for a cluster of rooms, entirely devoid of supplementary protective devices.

No wells for water supply have been detected.

The configuration of the Örentepe settlement, consisting of two nearly homologous levels excavated into the hillside, is unusual compared to the Cappadocian underground shelters, which typically extend predominantly on a single horizontal plane at the elevation of the terrain.

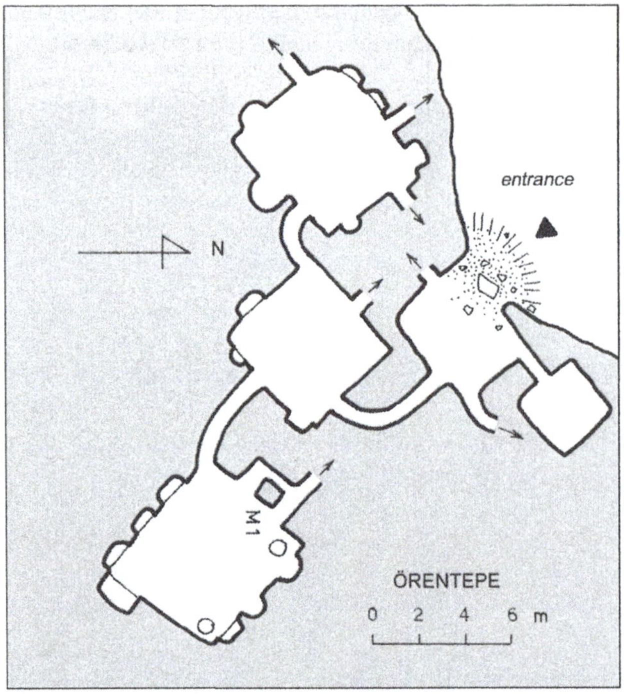

Fig. 4.58 Örentepe Underground Shelter plan (Bixio, 2012)

Örentepe appears to embody a transitional form between underground shelters and subterranean cities, such as Derinkuyu, situated approximately 18 km to the east.

In this context, more so than in other locations, the hypothesis that the construction of underground structures served as thermal refuges to protect against the harshness of the local climate appears justified. The average internal temperature, approximately 12 °C, which is considerably higher than that of other examined cavities, is ascribed to the exceptionally well-exposed and markedly darker rock.

Currently, it is impossible to evaluate the construction timelines for this site based on the existing research. As the construction of the underground structure appears more rudimentary than others, particularly in comparison to defense systems that are less numerous and less organized than in other locations, it may be older.

This aligns with Triolet's perspectives regarding the establishment of Derinkuyu, which appears to share certain similarities with Örentepe. Also, as a general principle, each Cappadocian underground shelter corresponds to a surface village. In the instance of Örentepe, there are no residences within a 3 km radius in any direction.

Consequently, additional archaeological inquiry is required to ascertain the precise characteristics of these circular areas and their potential connection to the subterranean structure unearthed above.

4.40 Ötedere Valley Underground Shelter No 1

The valley extending southward from Belağası Village, situated 7 km east of Kayseri, is known as the Ötedere Valley. Despite the absence of contemporary human habitation, the valley contains two subterranean shelters. The first shelter, named "Ötedere Valley Underground Shelter No 1," is situated at a distance of one kilometer from the valley entrance, on a precipitous rocky outcrop. This structure, which the local population refers to as the "Leprosy Hospital," is arguably one of the most intriguing structures in Kayseri. The structure has numerous rooms with broad windows facing the valley and stone masonry walls reaching a height of approximately 4–5 m (Fig. 4.59).

The presence of tunnels within the structure and two large millstone doors securing these tunnels provide

Fig. 4.59 Ötedere Valley Underground Shelter No 1 (photo A.E. Keskin)

evidence that the initial construction purpose of this structure was not a leprosy hospital, despite the presence of high stone walls defending the slope. Two tunnels extend in an eastward and westward direction from the room at the structure's entrance. The eastern tunnel is obstructed by debris at a point approximately three meters from the entrance. A millstone door shaft is located at the outset of the gallery, and it is evident that the eastern tunnel persists beyond the collapsed section. After a distance of 3.5 m in a westerly direction, a millstone door with a diameter of 1.20 m and a thickness of 26 cm, which was used for closing the remainder of the tunnel, remains in situ. The main gallery is defended by a millstone door and its intricate tunnels, which were left unfinished at a point 40 m below. Conversely, a minor branch of the aforementioned gallery extends to multiple rooms with windows that open to Ötedere Valley. Subsequently, these rooms lead to the largest chamber of the subterranean structure, which is connected to another lengthy tunnel.

Another tunnel, oriented eastward from this chamber, was constructed with significant engineering expertise. It features a 43° slope and a staircase that reaches another chamber on the top floor. The façade of the upper chamber is entirely open to Ötedere Valley (Fig. 4.60).

The structure does not resemble any other known underground shelter with this interesting architecture. Although rooms completely open to the valley and with broad windows give rise to the thought that it is a rocky settlement, there are also underground passages protected by millstone doors and a highly large chamber again excavated completely underground and supported with a column. When taken into consideration as a whole, it can be thought that this is an underground shelter that had been started to be excavated and left unfinished. Afterwards, some parts of the structure were used as cave settlements. The construction reason of the stone walls with 4–5 m height and surrounding this structure is unknown (Yamaç & Tok, 2015b).

4.41 Ötedere Valley Underground Shelter No 2

The shelter is situated on the eastern slope of Ötedere Valley. The site also features the carved-in-stone graves, situated in close proximity to the entrance of the shelter. Of particular interest is the millstone door, which is located outside the entrance, in the vicinity of the graves. It can be reasonably deduced that the millstone door in question

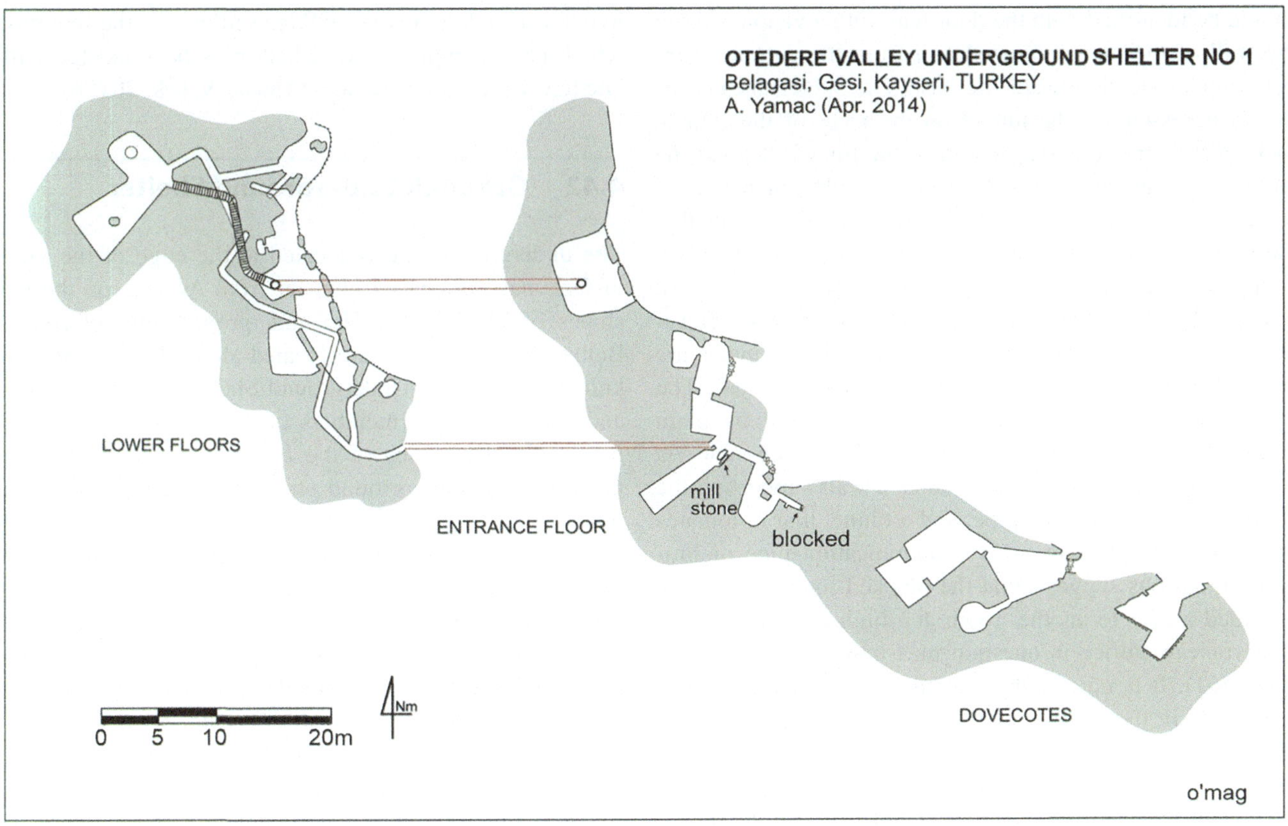

Fig. 4.60 Ötedere Valley Underground Shelter No 1 plan (drawn A. Yamaç)

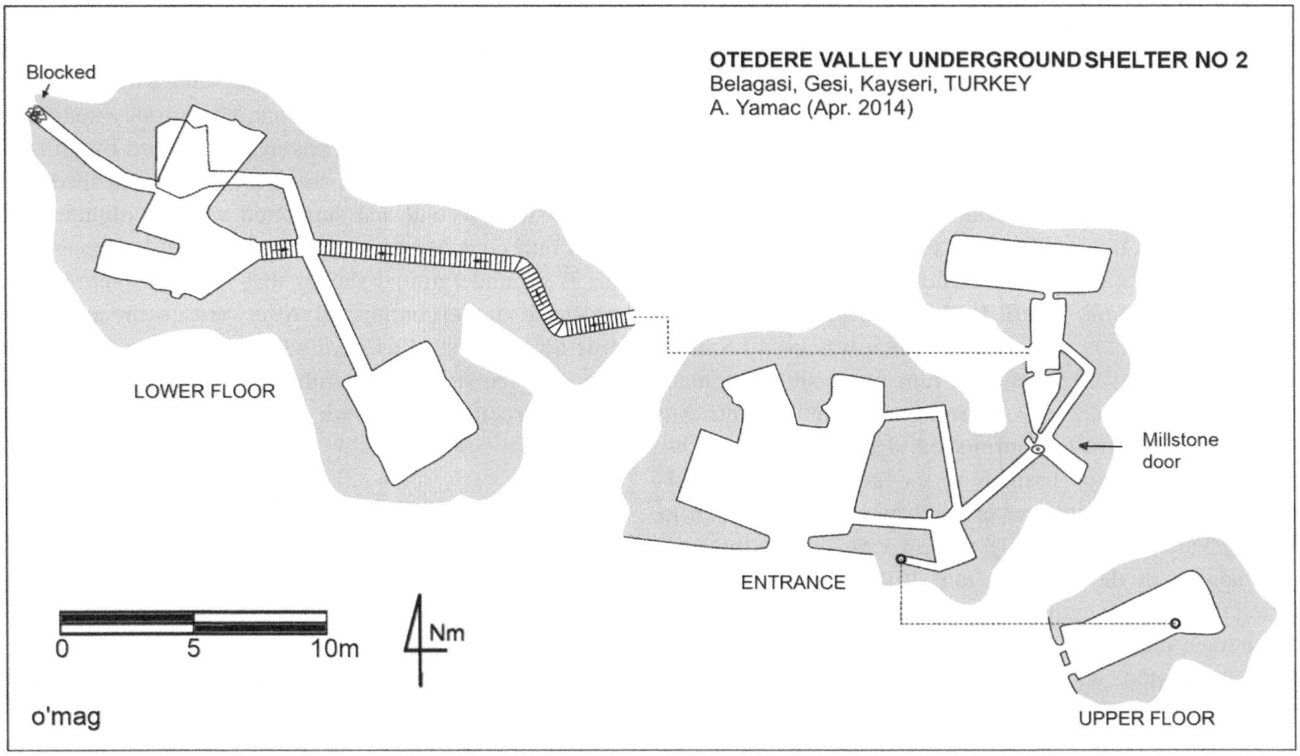

Fig. 4.61 Ötedere Valley Underground Shelter No 2 plan (drawn A. Yamaç)

should be identified with the door featuring a visible slot on the wall of the tunnel situated 4 m from the entrance tunnel, which extends toward the east. The subterranean edifice is accessed via the tunnel on the right of the sizable rock-carved structure. At the end of the tunnel, beyond the large in situ millstone door, a chamber with dimensions of 6 × 2 m is revealed. Of particular interest is the fact that the tunnel bends to the east before reaching this chamber, continuing for a distance of 13 m with a slope angle of 30. The depth of the tunnel is 6 m from the entrance elevation, where it terminates at a crossroads. Three tunnels are situated in this area, each leading to a separate hall. The northwest and southeast halls are characterized by plain architectural features, whereas the third hall, accessible via a short staircase in the west, features a wall supporting the ceiling, a communication pipe, and a tunnel that terminates after 8 m. The presence of a communication pipe of limited dimensions suggests that the blocked tunnel may have provided access to another room at a higher elevation. The subterranean edifice in question measures 160 m in length (Fig. 4.61). It is noteworthy that there is no evidence of a settlement in the vicinity of this underground shelter. Two underground shelters were constructed for defensive purposes at the exit of the valley and Belağası Village, which is situated 1 km away. It is therefore unlikely that the inhabitants of this settlement would have used this underground shelter in the valley in the event of an attack. An alternative hypothesis is that a small village settlement, the remains of which have completely vanished, may have existed within Ötedere Valley centuries ago (Yamaç & Tok, 2015b).

4.42 Özkonak Underground Shelter

The underground site is located at the edge of the village of Özkonak which is 14 km far from Avanos, on the north slopes of Mount Idis. The underground sites of Ayazma, Belha, and Yallı Damı are located along the way. It is also known as Genezin Underground Shelter. It was researched and mapped by the members of the National Commission for Artificial Cavities of the Italian Speleological Society in 2000. The underground structure is partly open to the public.

The underground network developed beneath a flat surface on which a lot of ventilation chimneys stand. Entry [1] is open on the edge of the tufa deposit, where it forms a step. From here we follow a series of chambers, more or less parallel to the small outside cliff, which is overlooked by other entrances, now closed with stone walls. The entrance hall [1] has two blocked tunnels, slightly downhill, which could lead to an unexplored sector.

From the room [7] a descending corridor starts, perpendicular to the cliff, which leads to an inner room [8]. From here a long corridor crosses two rooms [9] and [10] and reaches the room [11]. On the floor opens up a narrow tunnel, just buried, which could give access to hypothetical lower levels.

The site can be considered divided into two sectors. The first sector consists of the rooms numbered from [1] to [7a], is dug in the first band of the deposit of tufa rock. His disposition seems to respond to a predominantly residential use and it is not protected by millstone doors. Here there are many niches to store tools or supplies. In one of the rooms there is a large jar, still unbroken, but you can see many scattered shards. They have identified two places where the grapes were pressed, equipped with tanks for the collection of wine, communicating by means of little pipes.

The community, in an emergency, could find shelter in the second sector. In particular, the spaces [8] and [9] constituted a "redoubt," i.e., an internal area of refuge, protected by closure devices (M1, M2, M3) with millstone doors (Fig. 4.62).

The position of the point located in the M3 is to be checked carefully. Looking at the plan it should be obvious that the closing device would defend against possible raids coming from the innermost part of the network. Therefore, it is logical to think that the filled tunnel in the room [11], although today has more the appearance of a drained well, originally linked up with another system of underground spaces, in turn equipped with communications with the outside, independent from the present entrance (Fig. 4.63).

Very interesting is the monolith M2. This millstone door, of very remarkable dimensions, (180 cm in diameter and 50 cm thick) and without central hole, is positioned to be operated in a trench (locking parapet), rather than by means of pillars or locking slabs.

Moreover, the protection of the "redoubt" was reinforced by an active defense system consisting of a guard room above the connecting corridor [9a], that is reachable by a tunnel above the millstone door M2. Here two slots open on the ceiling allowed to hit any enemies intent to push or to make a hole in the millstone door, even without taking aim (Bixio, 2012; Gülyaz & Yenipınar, 2007).

4.43 Özlüce (Zile) Underground Shelter

The underground shelter is situated within the confines of Özlüce Village, Derinkuyu district of Nevşehir province. The village, previously named as Zile (the ancient Zeila), located on the slopes of Erdaş Mountain, is accessible via a 6 km road that diverges from Kaymaklı town on the Nevşehir-Derinkuyu highway.

During the rescue excavation and cleaning work conducted by the Nevşehir Museum Directorate in 1991, new galleries were unearthed (Şahin, 1993). The entrance to the Özlüce Underground Shelter is located to the east, and the main tuff rock is accessed via two interlocking arched spaces constructed from rubble stone and a 15-meter-long corridor, also made of rubble stone. A staircase constructed from stone leads from the corridor to the main area. The stone-built spaces that provide access to the underground shelter are considerably more recent than the rock-carved spaces that comprise the main underground shelter. At the end of the corridor, one encounters a millstone door of 1.75 cm in diameter, crafted from hard granite. The rock-carved space opens onto galleries on four sides. It should be noted that ventilation shafts and communication holes

Fig. 4.62 ÖzkonakUnderground Shelter (photo M. Gülyaz)

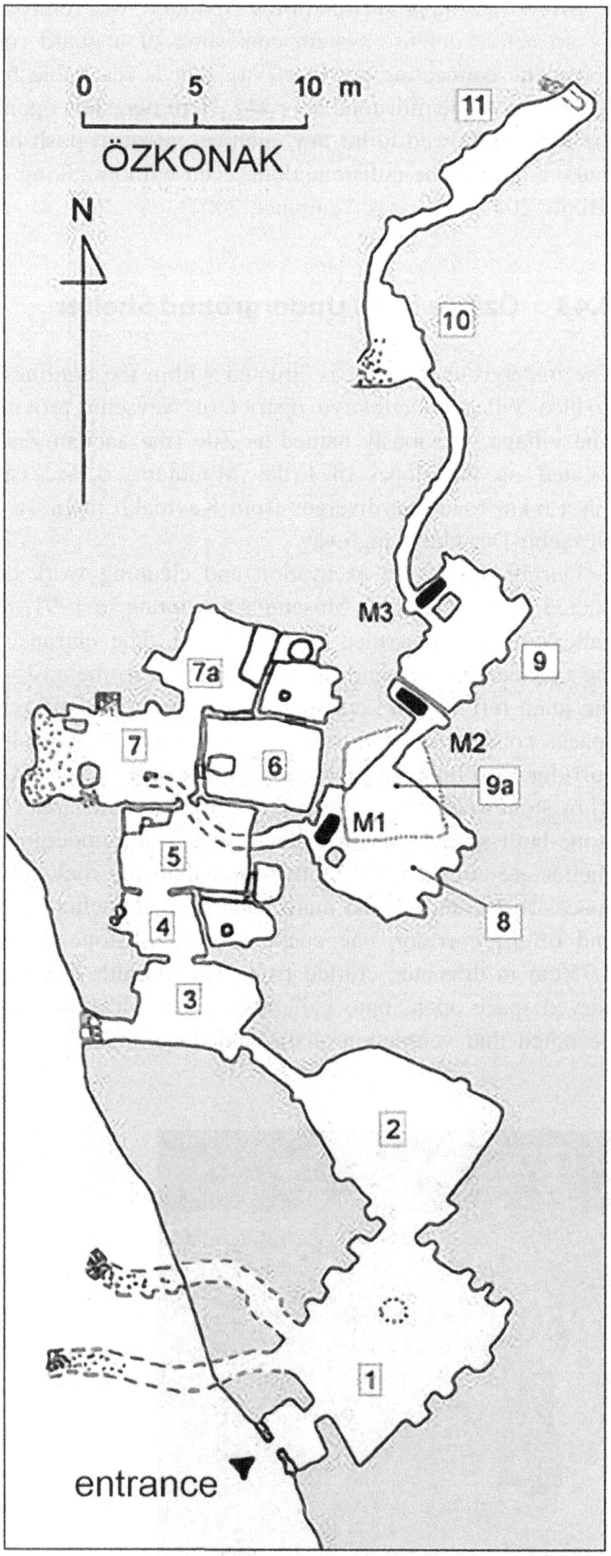

Fig. 4.63 ÖzkonakUnderground Shelter plan (Bixio, 2012)

are present in this area. The soil accumulation in the gallery number 3, which is formed by the main area, was cleaned. In the work carried out in the gallery number 4, which was opened to the north of this, a warehouse was revealed. In the work carried out here, a narrow corridor was encountered. Again, in the gallery number 5, which is located to the south of the main area, a ventilation shaft was detected and cleaned of the existing soil. Cleaning work was carried out in gallery number 6. During the work, two separate corridors were revealed in the south and east directions. Work was intensified in the corridor in the east direction. As is the case with other underground cities, each of the aforementioned corridors is connected to the surface dwellings. In periods of conflict and threat, individuals sought refuge in these corridors and the underground cities that they connected to. A ventilation shaft is located in gallery number 10. Furthermore, the excavation of this area revealed the existence of an additional corridor extending in a northwest direction.

Gallery number 11 was reached through a curved and human-height corridor. This gallery, which was completely full, was partially cleared. In the cleared area, a narrow corridor of 90 cm height and 70 cm width was revealed toward the west. The small galleries numbered 13–14 and 15 on the very long and curved corridor were probably built for resting purposes, while the wells with a diameter of 70 cm were probably built for trap purposes. Gallery numbered 16 is larger, and a well with a diameter of 1.5 m was found here. The corridor, which continues westward from this area in a curve, continues up to the hill next to the village. This length is approximately 500 m.

The underground settlement is partially flooded with sewage from the houses above. The site develops on a single horizontal plane. Except in some variations, it appears to correspond to the type of an "underground shelter," consisting of an area on the first band (rooms [2] and [3]), linked to an inner redoubt [4]. The long tunnel could be the connection between two equivalent units (the second of which is currently inaccessible). The hall [2] clearly had also the function of clearing node to other sectors of the settlement (Fig. 4.64).

4.44 Pazarören Underground Shelter

The structure is located on a rocky hill approximately 3 km south of Pazarören Village, within the Pınarbaşı district, positioned to the southwest of Kayseri. This historical site, known as the Pazarören Shelter, has been officially

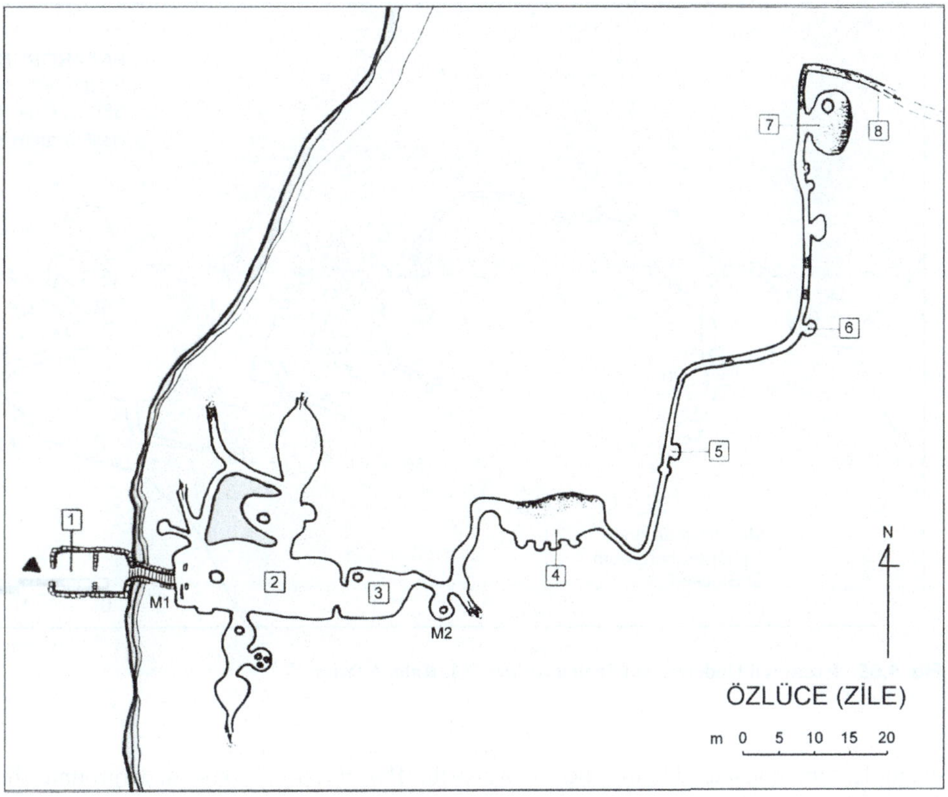

Fig. 4.64 Özlüce (Zile) Underground Shelter plan (Bixio, 2012)

designated as a First Degree Archaeological Protection Site due to its significant historical and architectural value. For a period of time, efforts were made to open this underground shelter to tourism, with the aim of making it accessible to visitors and researchers alike. However, these attempts ultimately proved unsuccessful, and the site was eventually abandoned, left to deteriorate over time without further preservation efforts.

Despite the challenges posed by debris obstructing certain passageways, which prevented complete measurements, the underground shelter remains one of the most substantial defensive structures in the region. Based on available measurements, it extends to an impressive total length of 420 m. The structure is uniquely positioned, occupying the entire base of a small hill and featuring multiple entrances situated on both its slopes. These entrances, along with the extensive underground layout, highlight the strategic nature of its construction.

The shelter consists of six operational rooms, of which only three are equipped with millstone doors. These heavy, circular stone doors were a common defensive feature in underground structures, allowing inhabitants to seal off sections of the shelter in the event of an attack. Interestingly, the two wide entrances located on the northern side of the shelter do not align with the traditional design of a defensive structure. This suggests that these two galleries were most likely added at a later period, after the site had lost its original defensive function. With the exception of these additional spaces, the underground shelter primarily comprises four substantial chambers. Notably, the two central chambers stand out due to their sheer size, each exceeding 150 m^2 in area, making them among the largest rooms within the complex (Fig. 4.65).

Two tunnels extend from the shelter, one leading to the west and the other to the east, though both are currently blocked by debris. Within the west tunnel, an operational room is positioned in a way that suggests it was designed for defensive purposes, providing protection against potential threats approaching from that direction. However, this particular room does not feature a millstone door, unlike some of the other rooms within the structure. Despite extensive surface research and investigation, the original entrance to this westward-extending tunnel has not been identified, leaving some aspects of the site's layout and function still open to further exploration.

One particularly notable aspect of the Pazarören Shelter is its geographical placement. It is situated approximately 3 km south of Pazarören Village, a location that raises intriguing questions about its relationship with surrounding settlements. Given that similar underground shelters built for defensive purposes were often positioned in close proximity to or even directly beneath inhabited areas, it is reasonable to assume that a settlement may have once existed near this structure. In comparison to other known

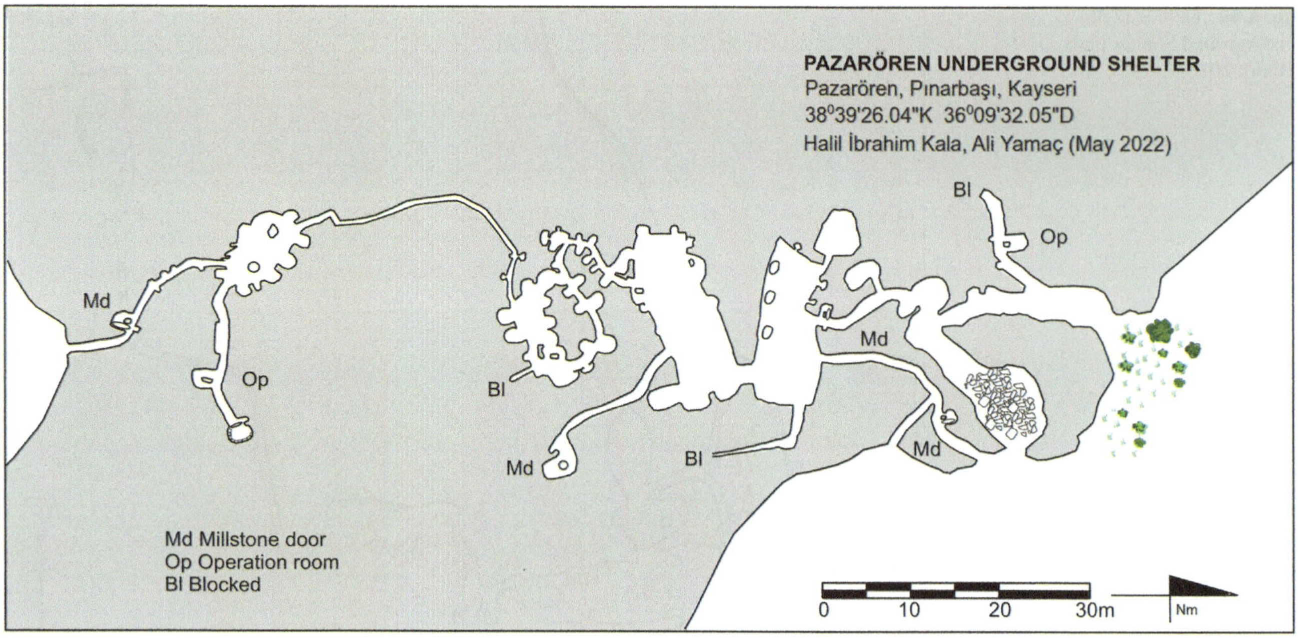

Fig. 4.65 Pazarören Underground Shelter (drawn H.İ. Kala, A. Yamaç)

examples, the considerable distance between the Pazarören Underground Shelter and the present-day village suggests the possible existence of an additional settlement during the Middle Ages. This settlement, which may have been located near or even above the underground structure, could have been abandoned or destroyed over time, leaving little trace of its presence.

As with many similar historical sites in the region, the Pazarören Shelter provides valuable insight into the defensive strategies and architectural ingenuity of past civilizations. Its extensive network of rooms and tunnels, coupled with its strategic location, underscores the complexity of its original purpose. Though currently left in a state of neglect, the shelter remains an important part of the historical and archaeological landscape of the Kayseri region.

4.45 Sivasa Underground Shelter

The underground shelter is located near Gökçetoprak Village, 35 km west of Gülsehir District. The studies were initiated by a French research team in 1989 and it was researched and mapped by the members of the National Commission for Artificial Cavities of the Italian Speleological Society in 1999. The village is set amidst a natural amphitheater of the cliff that forms the eastern side of a small butte. The buildings have a very archaic aspect: most are built in stone walls and seem to blend into the steps of the cliff with which they are partly fused.

The underground shelter of Sivasa has many similarities with those of Göstesin and Filiktepe, near the village of Ovaören, located 2.5 km to the south. The underground network was dug inside one of the modest elevations of the ignimbrite, almost elliptical in shape, with the major axis north–south, scattered throughout the extensive alluvial plain bounded on the north by the course of Kızılırmak River.

The shelter consists of a series of systems, all the entrances of which are open, on the side of the slope facing to the east. Seven of them have been identified so far and probably there are more. The entrances spread over the front area of about 1.200 m. Some of them were closed by the doors and stone doors. Some parts of these doors bear obsidian or sharpened flint stones that were used for harvesting grain on their lower sections. Except the north area containing underground rooms that were excavated parallel to the profile of the hill, it seems like there are no underground rooms in the other parts of Taniktepe. One of these rooms could be a church. Some of the paintings on the half dome are barely visible. Only S7 and S8 of the seven systems in the east section are linked to each other. It is believed that the other systems are independent. However, there could be more linking tunnels that couldn't be identified due to the secrecy of the local people and the natural obstacles encountered during the investigation.

System S1: The underground structure has been damaged due to collapse. Therefore, the original size of the structure is not known. The results of the current studies show that S1 covers an area of about 420 m of two layers and it is the most complicated system of this underground shelter. The system consists of two sectors that are almost alike. The sectors are only accessible through the entrance S1/a, opening to the isolated room that is protected by the millstone doors M1, M2, and M14. Both sectors consist of a series of sections with many independent exits that were aligned on the same level with the entrance. Each sector has a tunnel heading to the lower floors. These floors comprise several rooms opening to the tunnels that have been blocked due to landslip and collapse. A great number of stone door equipments, 26 of which could be identified, have been discovered so far in the region (these doors are designated by the symbol of M and a number). The dispersion of these doors jibes with the complex of the underground system (Fig. 4.66).

System S2: Development about 180 m. It consists of rooms with a length of 30 m. The cells carved into the walls were probably used as feed mangers. The feed mangers verified that the Kiratli underground system was used as a stable.

System S4: It measures about 120 m. The long room found in System S2 was probably used as a stable. The function of the tunnel in the shape of an arch discovered in the upper section of the room is not known. The tunnel suddenly ends with rock. The other sections in the area haven't been investigated yet. The millstone doors near the entrance were removed from their original places.

System S6: It measures about 105 m. The millstone found in Area M2 is closed. After removing the fill of the operating room, it became a folding door and it can be opened and closed by two people.

System S7 (about 192 m) and S8 (about 108 m) consist of two units (S7a, S7b and S8a, S8b): The room protected by two closing equipment M5 and M6 is the gradient point of the system. Two millstone doors on both sides protect this isolated room that serves as the corner stone of two main systems. In Area M7, there is a freestanding stone door. This kind of precaution is rarely seen. A corridor that is longer than 50 m links the units of S7a and S7b to each other.

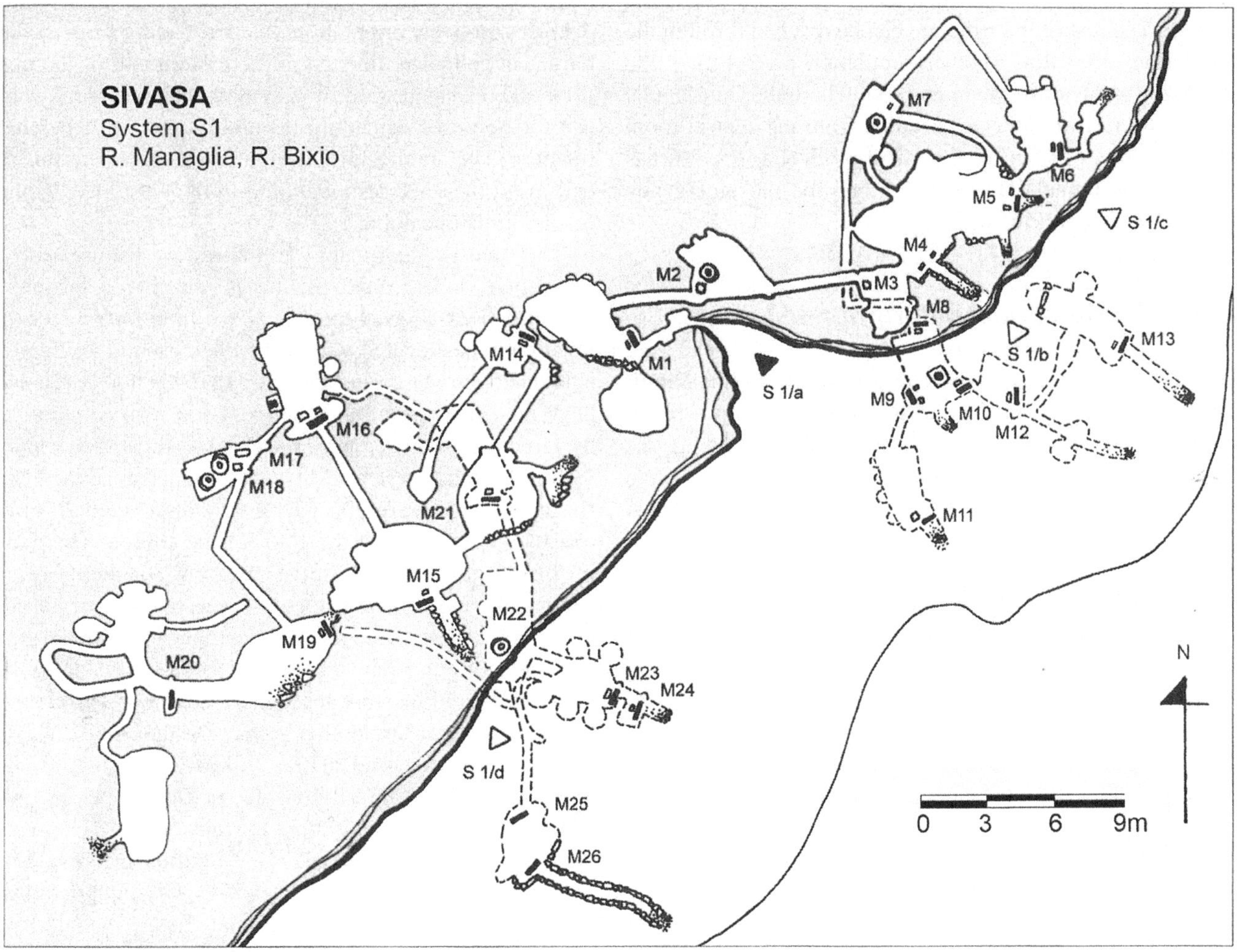

Fig. 4.66 Sivasa Underground Shelter System S1 plan (Bixio, 2012)

System S9 (about 174 m): It consists of 5 units located perpendicular to the slope. These units located parallel to each other are connected. The fact that the last two rooms are blocked suggests that there could be more units. These units were built to reinforce the defense of the units forming the system.

System S10 (about 210 m): It consists of two large twin rooms (S10a and S10b) connected to each other. The second room reaches to the inner section where several passages are located (at least six) and this section is blocked today. There is also a vertical pipe (P2) reaching to another room in the lower level in the room. It is believed that the room located in the area where the millstone doors M2 and M3 were found was used as "isolation room." The blocked tunnel that should be linked to outside or some other rooms that could not be reached yet may be used as "emergency exit." In the room S10a, a blind tunnel reaches to the ventilation pipe (P1) that is probably connected to two rooms in the upper levels. The millstone door at the end of two rooms was protecting the opposite section of the entrance.

System S11: It measures about 207 m. At the entrance, traces of the upper floor, which has partially vanished, can be seen. The rest of the structure can be reached through the ventilation pipe (P1). Another ventilation pipe (P2) in the inner section heads to the large room where the long tunnel begins. There is a tunnel descending from the central room and the debris of a series of vertical pipes (P4, P5, P6) has blocked this tunnel. This corridor may be the emergency exit of the system (Fig. 4.67).

4.46 St. Mercurius Underground Shelter

The shelter is situated in a central position within Saratlı Village, Aksaray Province, and is constructed into a rocky slope that extends to the north. The structure comprises three floors and four known entrances. The exterior of the structure comprises a series of stone masonry settlements, some of which are of considerable antiquity. All of the entrance tunnels are located within these settlements. This is a common feature in Cappadocia, which has a history of concealing underground cities and providing shelter during emergencies.

As can be observed in the plan of the first and second floors, at the lowest level, the main entrance, which faces the main road, begins with a tunnel in the rear wall of an inn from the Seljuk period (fourteenth century) (Fig. 4.68). The ceilings of all the surface buildings are barrel vault roofs, which suggests that they may be the ruins of structures dating from the fourteenth to fifteenth centuries. The entrance floor of the underground shelter continues in three different directions within the medieval buildings, accessed via doors protected by millstones. Four distinct tunnels extend toward the east and west, safeguarded by five distinct millstones within a 50-m radius. The millstone door of the furthest east tunnel remains closed. The rooms in this section are relatively irregular in shape; however, they become more rectangular in form as one proceeds along the west tunnel. In the central portion of the tunnel, a collapse of the ceiling has occurred. This tunnel, which extends to the north and provides a connection to the second floor, is linked to a series of large rooms that appear to have been utilized as a kitchen and storage area. The tunnels that extend beyond the aforementioned two rooms are obstructed.

In the eastern part of the same floor, there is a tomb that is the subject of a vow made by the local population. The tomb, which has a separate entrance and is frequently visited by the residents of Saratlı Village, is the final resting place of an Alewi senior. It is highly unusual for a Muslim tomb to be located within an underground shelter. The local population is unaware of the details surrounding this particular tomb.

In contrast to the uneven surfaces observed on the lower levels, the third floor exhibits a nearly flat surface and a distinct east–west orientation. A door leading from the subterranean church is situated at the eastern end of the upper floor and opens outward. It is evident that the church was in use for purposes extending beyond its period of defensive function. The architectural design lends support to this theory, as all the galleries extending to the west of the church feature millstone doors.

The most significant architectural feature of St. Mercurius Underground Shelter is the church located on the third floor (Figs. 4.69 and 4.70). In comparison to the other churches located within the underground shelters, the church of St. Mercurius Underground Shelter is relatively large in size. The ceiling is supported by two columns, and the structure features a large apse and a preparation room.

In 2010, the Aksaray Museum conducted salvage excavations at the site, during which they discovered 20 graves that had been dug into the floor of the church. The majority of the graves are those of children, but there are also adult male and female skeletons. The remains of an adult male and female were interred together in a single grave. The underground shelter was subsequently designated as Saint Mercurius in consequence of the discovery of a headless skeleton in another grave. Despite the fact that it is known that the head of this saint was severed, the date of his demise is 280 (Binon, 1937; De Le Haye, 1909), which is a relatively early date for this church. The practice of in-church graves is a common tradition in Cappadocia. Additionally, the Goreme Open Air Museum exhibits a number of graves from various churches, the majority of which are dated to the eleventh and thirteenth centuries.

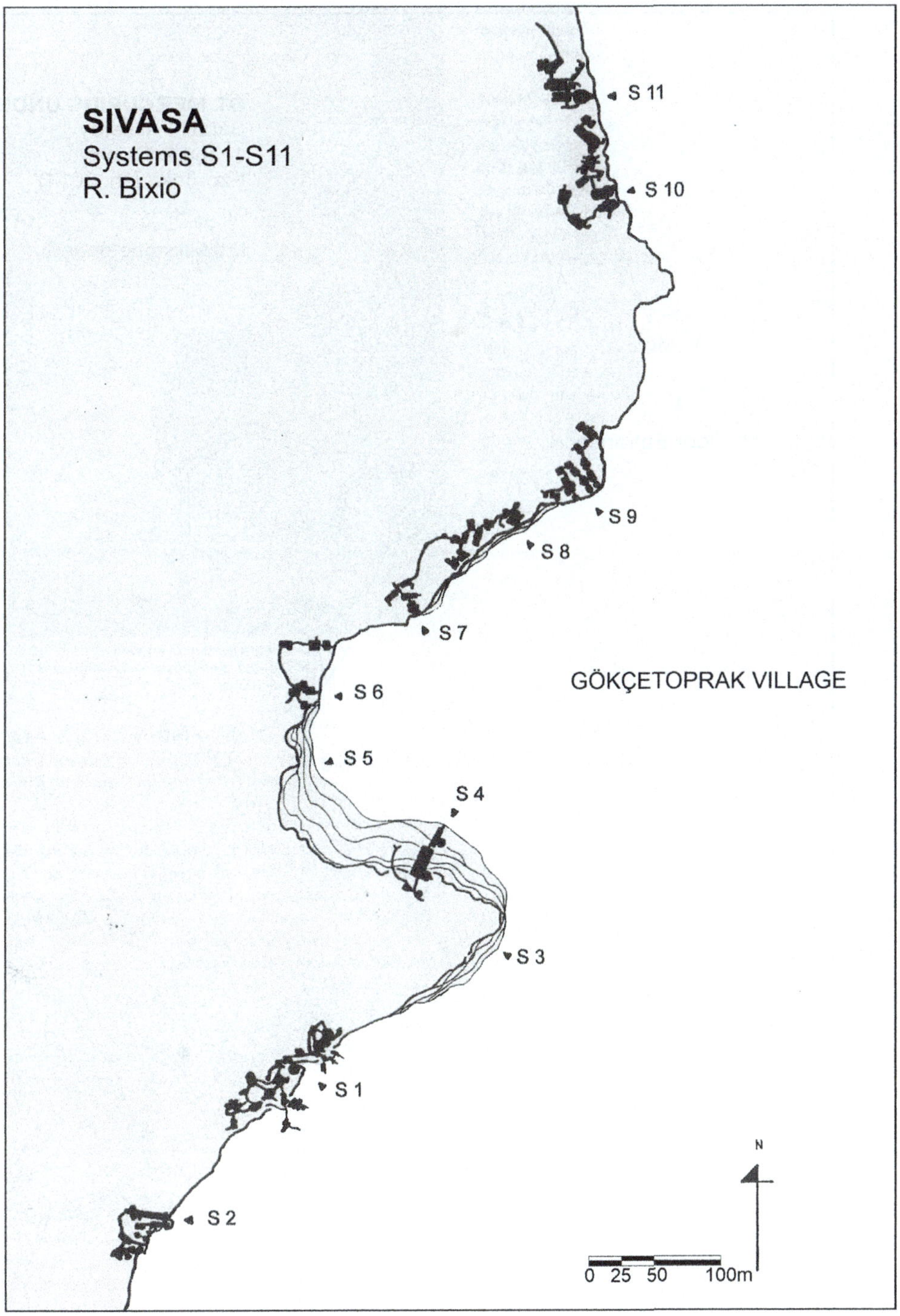

Fig. 4.67 Sivasa Underground Shelter plans of systems S1 to S11 (Bixio, 2012)

When the two side walls are considered as an "aisle," the long platform in front of the north wall of the church may be regarded as a kind of "stacidia." This architectural feature is not common in Cappadocian rock-dwelled churches. Additionally, the walls of the church feature carved crosses. The cross carved in a small niche in close proximity to the sanctuary is a type that has been observed in numerous locations throughout Cappadocia. This cross, designated the "Cross pattee," had been extensively utilized throughout the region.

The cross, designated as the "Cross pattee," was a prevalent form of ecclesiastical iconography in this region between the seventh and thirteenth centuries. The small room adjacent to the sanctuary, which contains seven graves on the floor, may be considered a kind of "preparation room." These architectural characteristics are typical of Eastern Orthodox churches. The dating of the church is challenging without solid evidence or C14 test results for the skeletons. However, a rough estimation suggests that the existing structure is likely to be no older than the seventh century.

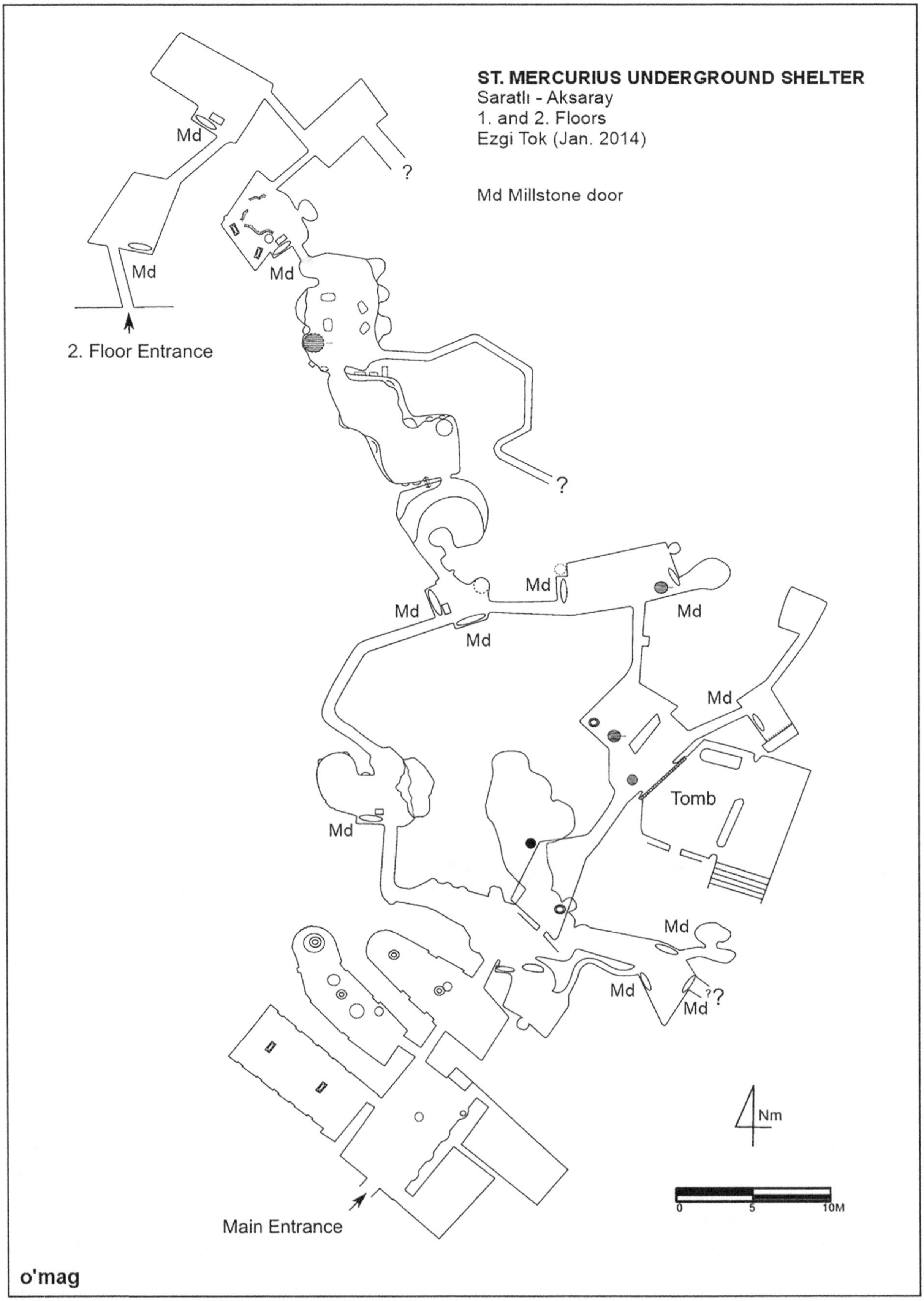

Fig. 4.68 St. Mercurius Underground Shelter 1. and 2. floors (drawn E. Tok)

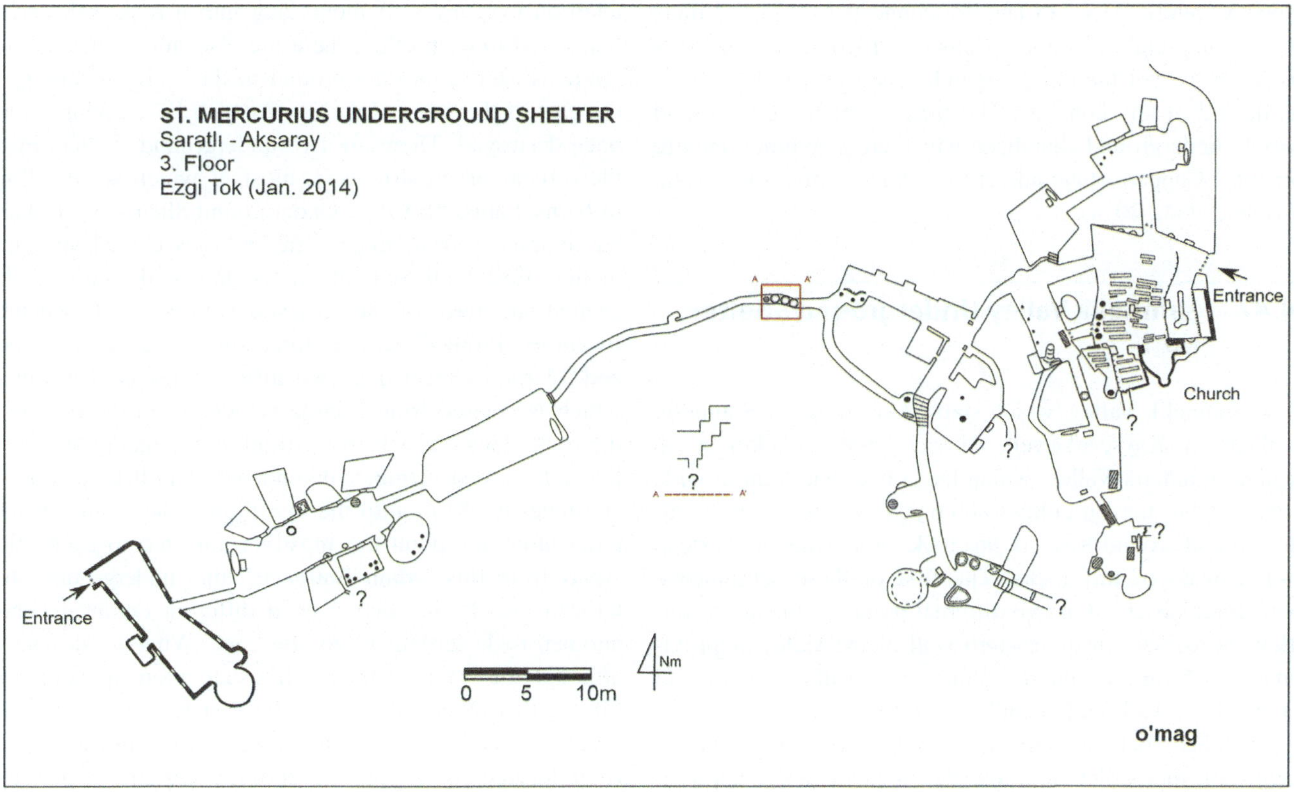

Fig. 4.69 St. Mercurius Underground Shelter 3. floor (drawn E. Tok)

Fig. 4.70 Church and the graves in St. Mercurius Underground Shelter (photo E. Soner)

Three of the four distinct galleries on the third floor remain inaccessible; one is obstructed by debris, and the other two are still equipped with closed millstone doors. The upper floor continues into two large rooms, one of which is a silo. It is possible that these blocked tunnels extend to another underground shelter, known as Kırkgöz, situated at a distance of 800 m. The western end of this floor reaches a stone masonry building with a barrel vault roof, which has been extracted from the site. The larger of the two chambers on this floor, which measure approximately 10 × 4 m and are connected by long galleries, contain eight distinct settlements or storage rooms. It is noteworthy that the other large chamber, measuring 48 m^2, is devoid of any contents. The function of these large and empty underground chambers, which are a common feature of some Cappadocian underground cities, remains uncertain (Yamaç et al., 2014).

4.47 Şahmelik Valley Underground Shelter No 1

The Şahmelik Valley, which starts 2 km north of Şahmelik village in Kayseri/Develi District, extends 6 km north again. Şahmelik Valley, which has a bed that is quite wide and suitable for agriculture compared to other narrow valleys found around Kayseri, has rock settlements on its steep walls made of tuff rocks. The first of these settlements, which are observed at two different points, is the dense settlement located on the eastern wall of the valley, approximately 3.5 km after the beginning of the valley, and we call it the "East Wall Settlement" (Fig. 4.71).

Another rock settlement located at the northernmost point of the valley and on the western wall, approximately 2 km after these structures, is called the "West Wall Settlement" by us. The East Wall Settlement, which consists of numerous interconnected rock structures, was a settlement in the early construction periods, but it has been partially converted into dovecotes in the centuries that have passed and most probably after it was abandoned. In addition to many settlements dug into a rock wall and all connected to each other, there are also numerous rock settlements carved into the ground in the valley extending to the east of the same rock, all of which have collapsed and been destroyed. There are two underground shelters inside these rock-carved structures, most of which are dwellings or barns. Şahmelik Valley Underground Shelter No 1, which has entrances from three different rock-carved structures in the "East Wall Settlement," is one of the large underground structures of the "Kayseri Underground Structures Inventory Project" with a total survey length of 468 m and 12 rooms located at two different levels. The tunnel, which is entered from a large settlement in the middle of the rock block and which we think is the main entrance of this defensive structure, is defended by a millstone door and continues for 82 m until the first room. The rooms encountered after this point are mostly connected to each other. Apart from this "Main Entrance," this underground shelter can also be reached from a different entrance behind another rock settlement to the east. What is strange is, although the "Main Entrance" has such a long and defended tunnel, this other entrance in the east has no defense and reaches directly to the inner rooms. In contrast, another room located on the upper level of this entrance is protected by a millstone door. This upper room in question is completely filled with alluvium. There is a third entrance to the underground shelter to the west of the main entrance. The beginning of this entrance is a settlement whose ground is also filled with alluvium. The tunnel continuing from this room to the west is blocked. The eastern tunnel makes a

Fig. 4.71 Şahmelik Valley East Wall (photo A. Yamaç)

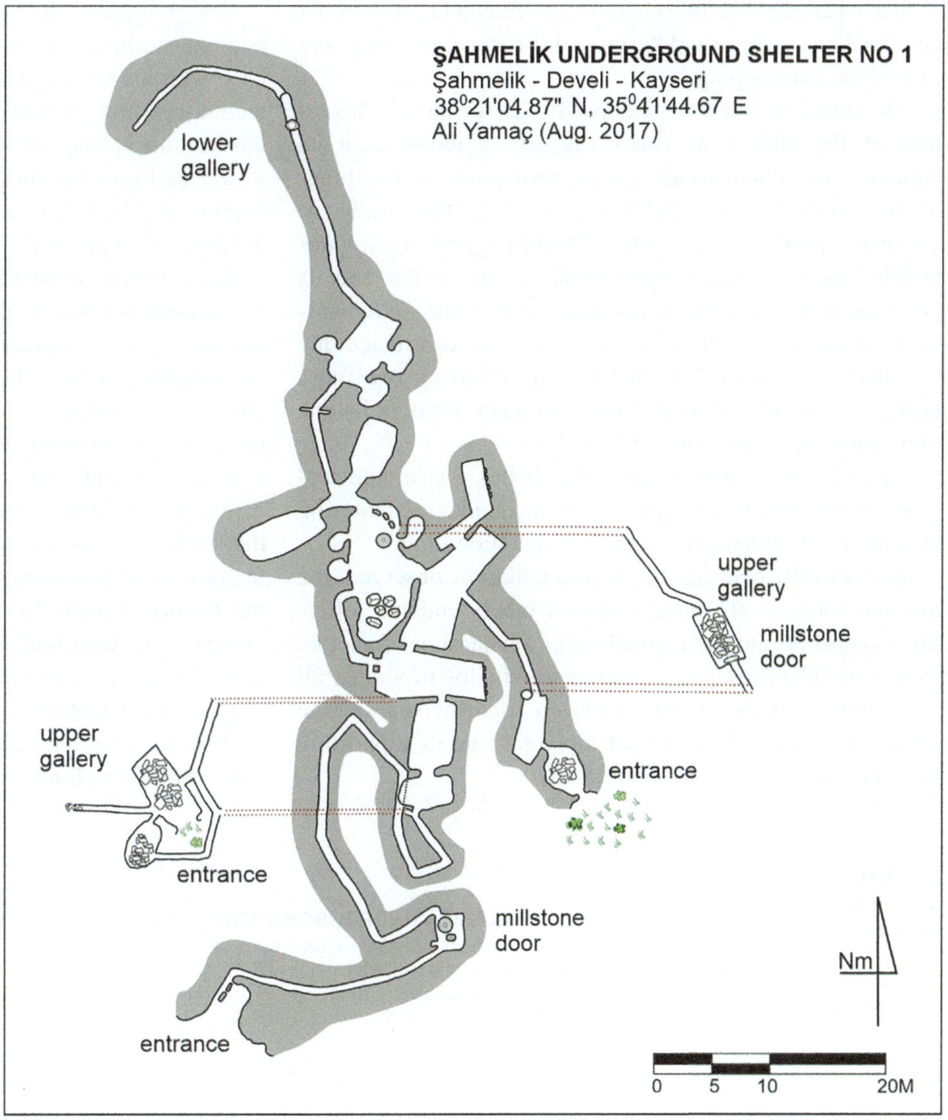

Fig. 4.72 Şahmelik Valley Underground Shelter No 1 plan (drawn A. Yamaç)

fork and reaches the inner rooms from two different points. The large millstone door on the floor of the north hall must have been moved here from the operations room in the south hall. The underground shelter continues northward with three different rooms after these large halls. After the last room to the north, a 70 cm wide tunnel continues 43 m and ends (Fig. 4.72) (Yamaç, 2018a).

4.48 Talasıra Underground Shelter

Compared with several underground shelters of Kayseri Province, Talasıra Underground Shelter is smaller than most of them. On the other hand, located on a lowland between Güzelköy and Gürpınar villages 15 km east of Kayseri, this underground shelter is a nice example of a perfect architectural utilization of a simple and small natural formation for a social requirement in the ancient times. The only tuffaceous outcropping entirely surrounded by a ground cover on a flat lowland was dug and turned into an underground shelter. Geologically, the tuffaceous rock platform forming the foundation of Talasıra Underground Shelter suggests a unique formation suitable for excavation, indicative of the geological considerations influencing ancient subterranean architecture. The architectural utilization of natural formations for defensive structures, as exemplified by Talasıra, underscores the adaptive strategies employed by ancient communities in response to external threats. This round tuffaceous rock platform is approximately 300 m in diameter and was completely undermined in the southeast section to form a defense structure that is interconnected by long tunnels.

When constructed, the builders of the structure kept away from relatively broken and weak rocks in the west section of the rock block. Indeed, the west section of the shelter where the last two chambers were dug has a granular and weaker structure unlike the solid rock in the east section of the outcropping.

The adaptive reuse of collapsed areas in the south section of the shelter as barns suggests a multifunctional approach to subterranean spaces, indicative of the pragmatic utilization of available resources. The study of Talasıra's main room, extending 16 m to the north, provides insight into the spatial organization of the shelter and its potential significance as a communal space. The settlements in the north section of outcropping are fully destructed due to collapse of rocks. The shelter is accessed by two more tunnels from that section; both entrances have millstone doors and operation rooms (Fig. 4.73).

Apart from the north part of the shelter, the middle section adds another layer to the shelter's complexity, featuring an entrance with a millstone and an operation room.

Remarkably, beneath the surface collapses observed, the original form of all tunnels and chambers endures, offering a snapshot into the meticulous craftsmanship of ancient architects. The presence of underground silos in almost all 12 chambers of the shelter implies a sophisticated storage system that could have played a pivotal role in supporting the community during times of crisis.

The intriguing half-circle tunnel observed in Talasıra and its similarity to the one in Derinkuyu Underground City warrant further investigation into shared construction techniques and purposes across distinct sites. Those half-circle tunnels may be considered to have been excavated to dig and remove out the main rock in the center and to expand the hall. The existence of six millstone doors in Talasıra Underground Shelter signifies a multi-layered security system, reinforcing the importance of safeguarding the community within the shelter. The dislocation of some millstones and the preservation of others in situ add layers of complexity to the shelter's history (Fig. 4.74). The presence of additional cave dwellings to the south of the shelter, some abandoned and partially collapsed, indicates a dynamic cultural landscape surrounding Talasıra.

The identification of illegally excavated ancient graves in the south and west of the outcropping emphasizes the need for protective measures and conservation efforts to preserve the historical integrity of the site. A total of 22 excavated graves were detected, but it is likely that the number is much higher, prompting further anthropological and archaeological investigations.

This underground shelter must be the defense system of a small cliff-dwelling village that was excavated in the same

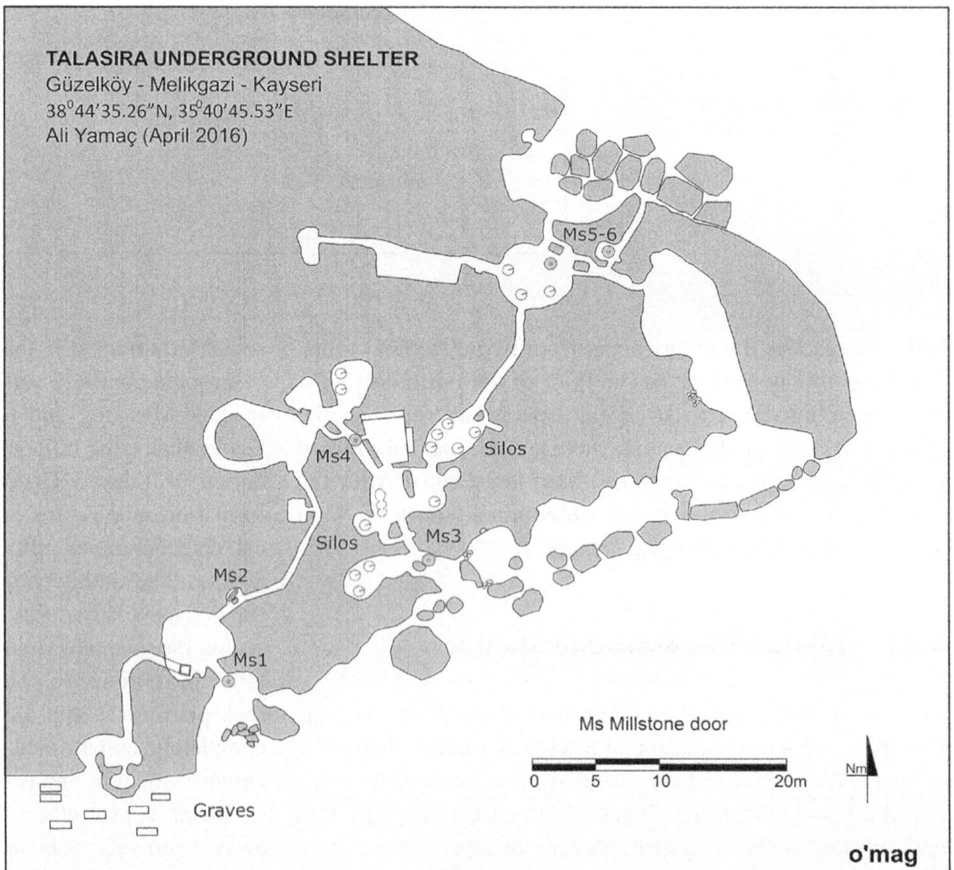

Fig. 4.73 Talasıra underground shelter plan (drawn A. Yamaç)

Fig. 4.74 Entrance of Talasıra Underground Shelter with a millstone door (photo A.E. Keskin)

outcropping, as it lies on a bare lowland with no ancient settlement nearby. A small population must have lived there, hidden in that shelter they dug to protect themselves during the raids and buried the dead into the graves located in front of the outcropping. The rock collapses in the north section are so large that it is not possible to make a comment either on the settlements of that section or on the total size of that cliff dwelled village.

4.49 Tatlarin Underground Shelter

The underground shelter is located on the rocky ridge in the northeast of Tatlarin Village, 10 km far from Acigöl District. The structure was initially researched by Sahin in 1993. Then, it was researched and mapped by several scholars. The underground shelter in Tatlarin was opened for visitors in 1991. The size of the rooms, the number of the storage houses, and the abundance of the churches indicate that the structure could be a military post or monastery complex. The underground shelter spread over a quite large area. Only a few parts of it have been cleaned. Two floors of it can still be visited. The most important feature of this underground shelter is that it has toilets and this feature is unique.

The underground shelter can be divided into three separate systems:

The first system consists of a series of partially underground rooms that were partly abandoned and integrated with several stone houses located mainly on the south slope of the village. The second and the third systems were directly carved into the straight walls of the slope.

The second system was carved into the southwest face of the slope overlooking Tatlarin Village. The debris of the rock wall unearths several sections of the system located at different elevations. Many of the underground rooms located in the area where the first group close to the surface is situated are quite narrow. Only the area in the center of the "slope village" was carved into the internal area. The entrance consists of a large rectangular room. The slightly inclined tunnel extends from the back wall and the corridor, which is reached after a few meters, is protected by the closed operating room. The millstone (this millstone is probably the one in front of the twin church) of this operating room was probably removed. The tunnel reaches the wide and long room. This room has a connection that is protected by a millstone door with double stone slabs. The opening, the right wall of which bears a column, reaches a small room below. This door, together with its additional section next to it, could be a toilet. Also, the entrance in this section provides a passage to the room that contains niches carved in the form of arcosolium, rocky platforms, and burial pits. The tunnel beginning from the opening reaches the rooms located on top of each other. The collapsed walls of these rooms are connected to the outside. This tunnel might have been used as "escape way" to evacuate the room in case of emergency or catch the enemy unawares. To the left of the room, the descending tunnel reaches the room at the lower elevation. This connection has two operating rooms with millstone doors that could be used to block the enemy. The operating rooms were built on the sections of the corridor that suddenly change. It is believed that this corridor was built to reinforce the defense system. A tunnel begins from the rooms and this tunnel is blocked due to collapse. It is seen that the tunnel is linked to the village. It is clear that two millstone doors in the operating room were placed to protect the room. The fact that the tunnel is only 60 cm high indicates that it was used as an "escape way." The tunnel opening to the first room and another room on the right extends toward the right and reaches the corridor, the operating room of which is protected by a millstone door. In the ground level, there are large open storehouses. These bell-shaped storehouses are plastered and were probably used for storing liquids. The room contains a millstone door that was found in the open operating room, and this columned door opens to the corridor.

The third system is separated from the other one but it is located on the same slope and situated a few meters north. The rooms were carved into pale pink colored, flat pumice facing west. The rooms are aligned on top of each other as five floors. It is hard to reach the uppermost sections because the stone walls of the passages connecting the rooms to each other have been damaged. The area consists of a series of rooms carved into the first arch of the slope and these rooms are connected to each other. In one of the upper floors, there is a tunnel reaching an unusually large gallery. It seems that this tunnel descends toward the inner sections of the mound. The further research will provide more information about the texture of this area (Figs. 4.75 and 4.76) (Bixio, 2012; Gülyaz & Yenipınar, 2007; Yamaç & Tok, 2015c).

4.50 Zemi Underground Shelter

The Zemi Underground Shelter is an artificial cave complex located in a standalone fairy chimney in Zemi Valley, near Göreme. Initially constructed as a utility space with elementary defensive functions, it was later adapted into a well-protected temporary refuge. Its development included the creation of hard-to-access upper-level spaces, ultimately repurposed into dovecotes.

This complex is among several in Cappadocia, whose study helps trace the evolution of cave construction in the region. Understanding its architectural features and underground spatial organization contributes to a future typology that may assist in dating and identifying different periods of Cappadocian cave usage.

Situated in a rocky cone on a natural terrace, the complex is surrounded by vineyards. It is 300 m northeast of Sarnych-Kilise and 450 m northeast of Karabulut-Kilise, with no other cave structures nearby. The entrance, positioned on the eastern slope, is accessed via a two-meter cliff with carved steps and hooks. Evidence of a former wooden door with frame cutouts and locking mechanisms remains. A narrow tunnel (1.5 m high, 0.8 m wide, 2.5 m long) leads to a fork, where a sliding millstone-shaped stone door, now overturned, once secured the entrance.

To the right of the fork lies Room 2, featuring a large grape press with a well-finished ceiling and walls. The back wall houses a wort receiver separated by a partition, with a drain and installation hole for a vessel. The room contains lamp niches, a rectangular shelf, and a neat window. To the left, a short passage leads to Room 1 (2.4 m high), which opens onto a cliff face. A narrow partition separates Room 1 from Chamber 1a, where feeders indicate the sheltering of livestock. Multiple utility niches are preserved, one featuring door cutouts and an internal lamp niche. Notably, two carved crosses and a smoothed wall section suggest religious significance.

In the northern wall of Room 1, an entrance at 1.5 m height leads to Well K1, connecting to the second tier. The well (4.25 m high, 0.8 m diameter) has step-like ledges carved for movement. The second tier opens into Room 3 (3 × 4 m, 2.2 m high), which has a rectangular doorway and access to surrounding chambers. Room 3a (1.5 × 1 m) and 3b (2 × 2 m) likely served as storage. Room 4 (over 2 m high) features a rectangular window and a niche with a

Fig. 4.75 Tatlarin Underground Shelter (photo M. Gülyaz)

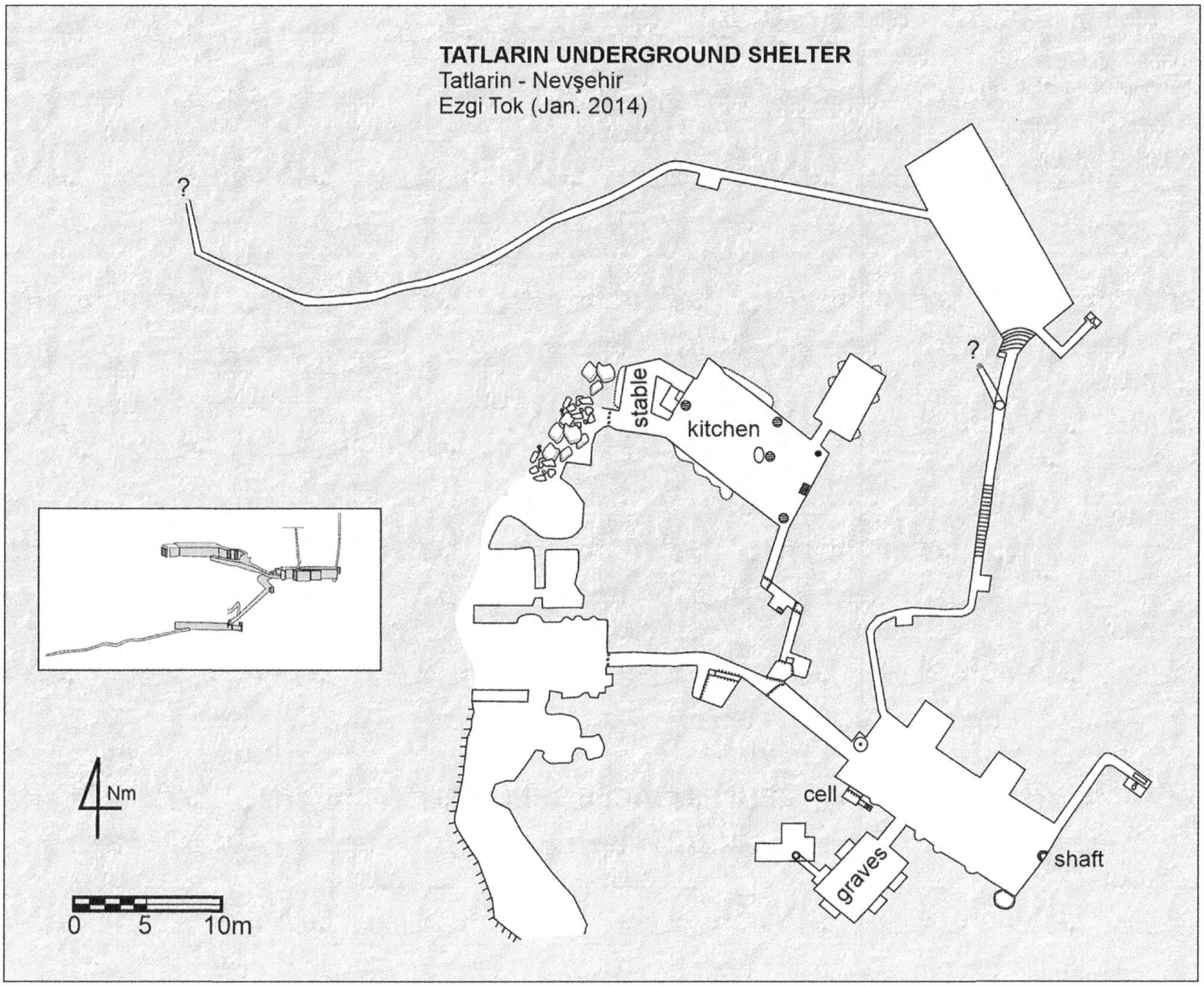

Fig. 4.76 Tatlarin Underground Shelter plan (drawn E. Tok)

carved cross, indicating use as a chapel. Crosses, inscribed in circles and painted in red ochre, adorn the walls. Traces of later expansion are evident, particularly in the destruction of a wall between Rooms 4 and 3b.

Room 5 (3 × 3 m, 1.6 m high) contains Well K3 (15.6 m deep), protected by a rim and featuring a groove for a hanging pulley. Steps lead to an external opening. Room 6 (2.2 m high) connects to Room 5 and displays three rows of bird-nest niches, dating to the dovecote phase. Room 7 (3 × 3 m, 2.1 m high) features a latrine and a water container, along with additional bird niches. From Room 5, a rock staircase leads to Well K2 (3.6 m high), providing access to the third tier. Room 8 (3.5 × 3.5 m, 2.1 m high) sits above, with openings to the rock face. Room 9, an extension, has a window, while Room 10 remains largely collapsed.

The complex's evolution indicates its initial use as a utility cave with defensive elements. The entrance, initially protected by a steep section and a wooden door, later included a millstone door. The presence of livestock feeders suggests an early function as a seasonal dwelling with agricultural and storage facilities. Modifications over time reinforced its defensive capacity, including the addition of Well K1, securing access to the second tier. The chapel room aligns with similar defensive cave shelters in Cappadocia. Later additions, including Chambers 6 and 7 as observation posts with latrines, further enhanced its protective functions.

As the structure evolved, third and fourth tiers were added, accessible only via Well K2. These upper rooms, now eroded, may have been reserved as a last line of defense. This pattern of unutilized upper-tier rooms is observed in similar Cappadocian shelters, such as in Meskendir Valley near Göreme and the Melendiz River Valley near Belisırma. Eventually, the complex's final stage of use involved conversion into a dovecote, evidenced by

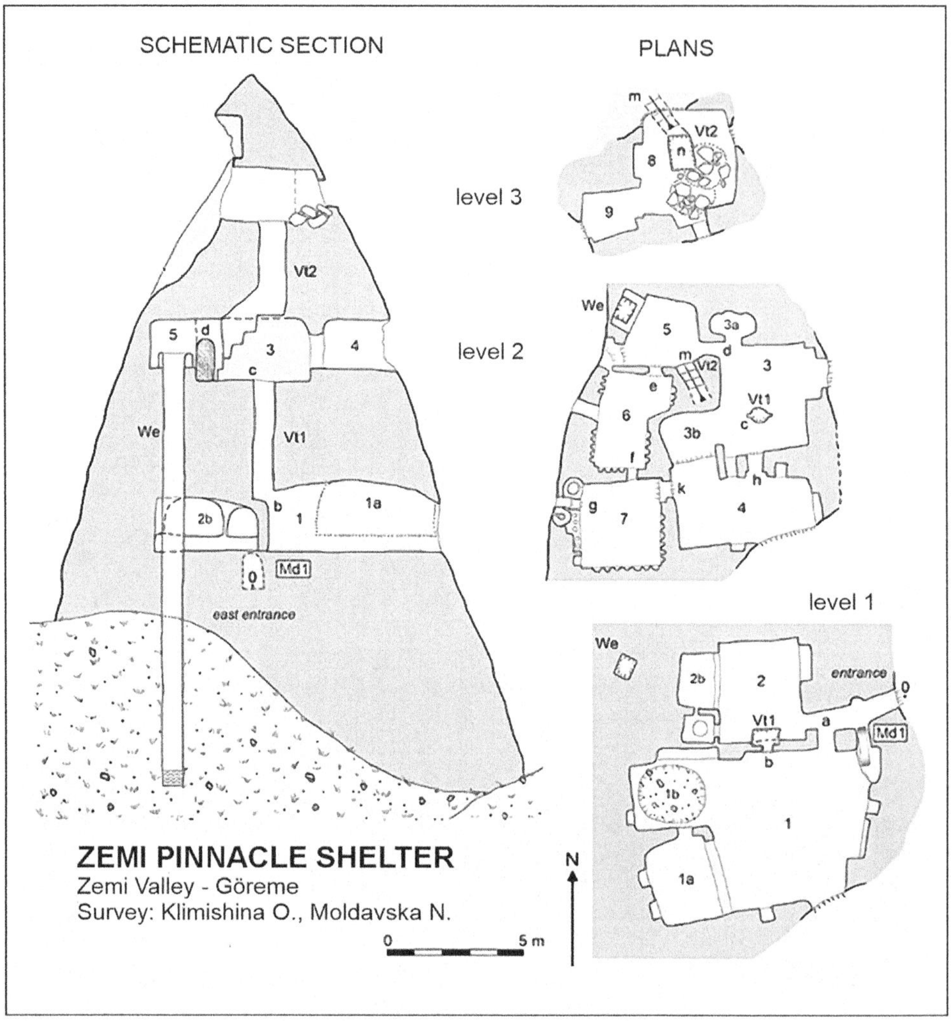

Fig. 4.77 Zemi Underground Shelter plan and profile (drawn O. Klimishina, N. Moldavska)

the numerous bird-nesting niches on the upper levels. This post-refuge phase is common among Cappadocia's hard-to-reach cave structures, indicating a practical repurposing of abandoned defensive shelters (Fig. 4.77) (Bobrovskyy et al., 2015).

References

Akok, M., & Gürçay, H. (1965). Yeraltı Şehirlerinde Bir İnceleme. *Türk Arkeoloji Dergisi, XIV*(1–2), 35–69.

Binon, S. (1937). *Documents Grecs Inédits Relatifs à St. Mercure De Césarée*. Bureaux du recueil, Bibliothèque de l'Université.

Bixio, R. (2012). *Cappadocia: Records of the underground sites*. Archaeopress.

Bobrovsky, T., Grek, I., & Moldavskaya, N. (2019). *Cave complex with a church and shelters in the Gomeda Valley (Central Cappadocia)*. Paper presented at the X International Scientific Conference 'Speleology and Spelestology', Nabereznye Chelny.

Bobrovskyy, T., & Grek, I. (2015). *An artificial cave complex in Meskendir Valley near Goreme (Cappadokia)*. Paper presented at the VI International Scientific Conference 'Speleology and Spelestology', Nabereznye Chelny. (in Russian)

Bobrovskyy, T., Grek, I., & Klimishina, O. (2015). *A complex in a standalone fairy chimney in Zemi Valley, Cappadokia*. Paper presented at the VI International Scientific Conference 'Speleology and Spelestology', Nabereznye Chelny. (in Russian)

Çakır, K., Savaş, F., Törk K. A., Acar, C., Kahraman, İ., Yeleser, L., Önal, S., Erşen, D., Cengiz, A. M., Tozsin, T., & Uygun, Ş. (2016). *Kırşehir İli Yeraltı Şehirleri Araştırma Raporu*. MTA Genel Müdürlüğü.

De Le Haye, H. (1909). *Les Légendes Grecques des Saints Militaires*. Librairie Alphonse.

Demir, Ö. (1990). *La Cappadocia, culla della storia*. Ajans Türk Publishing.

Dolotov, Y. (2014). New information about the cavities of the Cirali Obruk (Turkey). *Peshchery (Caves): Collection of Scientific Transactions* (37).

Giovannini, L. (1971). *Arts of Cappadocia*. Nagel.

Grek, I., & Podruckaya, N. (2012). *Cave City in Shiraly Lake Crater*. Paper presented at the III International Congress of Speleology and Speleostology, Naberznhye Chelny (Russia).

Gülyaz, M., & Yenipınar, H. (2007). *Underground cities of Cappadocia*. Dünya Kitap.

Hild, F., & Restle, M. (1981). *Kappadokien (Kappadokia, Charsianon, Sebasteia und Lykandos)*. Verlag der Österreichischen Akademie der Wissenshaften.

Inbaşı, M. (1993). *16. Yüzyıl Başlarında Kayseri*. İl Müdürlüğü Yayınları.

Kıvılcım, F. (2011). *A research on the conservation of the spirit of place (genius loci) Case study: Mustafapaşa (Sinasos) in Cappadocia* (PhD thesis). University of Rome, "Sapienza", Rome.

Kostof, S. (1972). *Caves of god: The monastic environment of Byzantine Cappadocia*. MIT Press.

Lucas, P. (2003). Les établissements monastiques de la basse vallée de Göreme et de ses abords. *Dossiers d'Archéologie, 283*(Mystérieuse Cappadoce), 32–41.

Nicoletti, M. (1980). *L'architettura delle caverne*. Editori Laterza.

OBRUK. (2017). *Kayseri Yeraltı Yapıları Envanteri* (Vol. 1). Kayseri Municipality.

Okuyucu, D. (2007). *Derinkuyu Yeraltı Şehri* (Derinkuyu Underground City) (Ma thesis). Atatürk Üniversitesi - Erzurum, Erzurum.

Ousterhout, R. (2005). *A Byzantine settlement in Cappadocia* (Vol. 42). Dumbarton Oaks.

Ousterhout, R. (2010). Remembering the dead in Byzantine Cappadocia: The architectural settings for commemoration. In O. Ioannisian & D. Jolshin (Eds.), *Architecture of Byzantium and Kievan Rus from the 9th to 12th centuries* (Vol. 53, pp. 89–100). Transactions of the State Hermitage Museum.

Ousterhout, R. (2017). *Visualizing community: Art, material culture, and settlement in Byzantine Cappadocia* (Vol. 46). Harvard University Press.

Özsoy, O. (2017). Gizemli Bir Menfez. *Şehir Kültür Sanat* (8), 8–17.

Öztürk, F. (2012). The unusual separation of Cappadocian refectories and kitchens: An enigma of architectural history. *METU JFA, 2012*(1), 153–169.

Popovic, S. (1998). The 'Trapeza' in Cenobitic Monasteries: Architectural and spiritual context. *Dumbarton Oaks Papers, 52*, 281–303.

Sözen, M. (1988). *Sinan, architect of ages: Arts in the age of Sinan*. Ministry of Culture and Tourism of the Turkish Republic.

Sözen, M. (2004). *Mimar Sinan Evi Restorasyon Projesi*. ÇEKÜL.

Şahin, S. (1993). *Özlüce Yeraltı Şehri*. Paper presented at the 3. Müze Kurtarma Kazıları Semineri, Ankara.

Yamaç, A. (2018a). *Underground shelters and Cliff dwellings of Şahmelik Valley (Kayseri – Turkey)*. Paper presented at the 4. MESS Symposium Antalya.

Yamaç, A. (2018b). Underground shelters of Tomarza (Kayseri – Turkey). *Opera Ipogea, 1*, 37–50.

Yamaç, A. (2019). Büyük Bürüngüz underground shelter (Kayseri – Turkey). *Opera Ipogea, 2*, 65–76.

Yamaç, A. (2022). Some interesting underground cities and peculiar underground structures of Kayseri (Turkey). *Jerusalem Journal of Archaeology* (3), 70–107.

Yamaç, A. (2023). *Rock-cut architecture and underground cities in Koramaz Valley of Kayseri, Turkey*. Springer AG.

Yamaç, A. (2024). Ağırnas: A monastery, a shelter or both? *Opera Ipogea, 2*.

Yamaç, A., & Gilli, E. (2016). Underground shelters of Golgoli (Cappadocia - Turkey). *Opera Ipogea, 1*, 33–40.

Yamaç, A., & Tok, E. (2015a). An architect's underground city. *Opera Ipogea, 1*, 37–46.

Yamaç, A., & Tok, E. (2015b). Cave dwellings and underground cities of Belagasi Village and Otedere Valley (Kayseri - Turkey). *Opera Ipogea, 2*, 41–50.

Yamaç, A., & Tok, E. (2015c). *Surveying some of the touristic underground cities of Cappadocia (Turkey)*. Paper presented at the HYPOGEA Rome.

Yamaç, A., & Tok, E. (2016). Doganli underground shelter (Capadocia - Turkey). *Opera Ipogea, 2*, 57–62.

Yamaç, A., Tok, E., & Filikci, B. (2014). St. Mercurius underground city of Saratli (Aksaray-Turkey). *Opera Ipogea, 2*, 37–46.

Yörükoğlu, Ö. (1989). *Underground cities in Cappadocia*. Aşik Ofset.

Yörükoğlu, Ö., Sevil, T., Taşcı, Z., & Türkmen, K. (1988). *Les Sites Souterrains de Cappadoce*. Aşik Ofset.

5 Conclusion

The structures analyzed in this book experienced one of the most tumultuous and violent epochs in Anatolia's history centuries ago. Beyond mere witnesses, they were the hope of survival for the inhabitants of the region during that period. Visiting a well-lit, easily accessible touristic underground shelter in Cappadocia today, one can imagine the individuals and families striving to hide from the plundering enemy outside in the same shelter, which was dimly lit by a few oil lamps 1300 years ago.

The millstone doors were shut, although screams and sounds from outside were audible. The enemy was forcing the doors to open. Inside the shelter, the air was heavy and food was scarce. They had been taking refuge here for three days, and it was unsure how many additional days they would remain, contingent upon the durability of the millstone doors.

Even millstones provided no assurance. Theophanis, in the eighth century, states: "In 776, during an incursion into Byzantine territory, the Arabs compelled a refuge named Kasin by smoking it out" (Theophanis, 1839). The narrative states that Byzantine refugees were apprehended and transported by Arabs under the command of the Abbasid Caliph al-Mahdi.

Cappadocia's underground shelters have witnessed countless such events over the centuries, and it is not without reason that they are the foremost example that comes to mind when subterranean shelters are discussed around the globe.

Protecting these shelters, which serve as significant historical and social emblems of that era, and how to pass them on to future generations presents a formidable challenge. Restoring and safeguarding all of these structures, already over 250, is nearly unfeasible, and the count continues to rise daily with new discoveries.

This issue is further compounded by the inability to adequately safeguard not only the rock-cut structures, such as shelters, dwellings, and cisterns, found in Cappadocia and left to their fate, but also the frescoed churches. Even these structures, which are the most important cultural treasures of Cappadocia and the main factor in its acceptance as a UNESCO World Heritage Site, are not protected as required.

The security and control of some churches in remote valleys have been entrusted to a local villager. Or a huge rock-cut complex such as Keşlik Monastery, south of Cemil village, is safeguarded by an individual asserting ownership of the edifice. Although palliative, these represent a form of solution, as in numerous other valleys one can enter a frescoed Byzantine church and explore freely; there are no guards, and some churches lack doors entirely.

The frescoes of the Church of the Forty Martyrs in Şahinefendi have undergone complete restoration as part of a long-term project led by Prof. Dr. Maria Andaloro from the University of Tuscia, involving over 30 experts (Andaloro, 2010; Andaloro et al., 2017). It is presently accessible without an attendant.

In this context, it is feasible to critique the accountable entities, specifically governmental bodies or municipalities. On the other hand, when the number of rock-cut churches alone in the region is more than 300, it becomes impractical for the authorities to devise a comprehensive solution for the protection of these structures. Ministry of Culture and Tourism, the exclusive authority in Turkey for this domain, lacks the personnel and financial resources to manage with so many antique structures.

Another solution in this situation is to hand over the renovation, restoration, and control of some structures to local governments or municipalities under the supervision of the General Directorate. The Ministry looks favorably on such projects and supports them. However, as there is no general consensus and project planning over this matter, and most of the local administrators who are the decision-makers in these activities are ignorant of these issues, not much progress can be made in this solution.

For example, as OBRUK Cave Research Group, we have been continuing our "Underground Structures Inventory Project" for 11 years and have so far discovered 41 underground shelters and 51 rock-cut churches in Kayseri

A. Yamaç and E. Tok, *Underground Shelters of Cappadocia, Turkey*, Advances in Karst Science,
https://doi.org/10.1007/978-3-032-16095-9_5

Province (Yamaç, 2021). All of these structures have been surveyed, inventoried, and preliminary reports have been submitted to the municipality. Notwithstanding our focus in private discussions beyond the reports, the municipality has failed to devise any conservation or tourism initiatives for these buildings, including the Büyük Bürüngüz Underground Shelter, one of the largest in Turkey (Yamaç, 2019).

This example illustrates that certain structures can be conserved by tourism; nevertheless, this necessitates that local authorities possess vision and expertise. The 2022–2024 annual average visitor numbers of some important historical places in the Cappadocia region are as follows (DÖSIM, 2025):

Göreme Open Air Museum	1.450.000
Kaymaklı Underground Shelter	630.000
Peristrema Valley	570.000
Derinkuyu Underground Shelter	460.000
Zelve Rock-Cut Settlement	280.000
Kırkgöz Underground Shelter	190.000
St. Mercurius Underground Shelter	140.000

Currently, Turkey has 17 underground shelters accessible to tourists, and with the exception of the four listed above, their visitor numbers are quite minimal.

Kırkgöz and St. Mercurius shelters exemplify effective preservation of these structures while also generating revenue. Both shelters are situated in Saratlı, a small village within the Aksaray province. The local municipality has renovated and offered these shelters for tourism. The revenue generated from ticket sales is allocated to the municipality and constitutes a significant portion of its budget.

Considering that most of the foreign tourists come to Cappadocia by travel agencies, it is easy to understand how advantageous the location of Saratlı is. It is located on the Aksaray-Nevşehir highway, and very close to the Peristrema Valley. On the other hand, Saratlı Municipality has been carrying out an intensive advertising campaign for years and gives promotions to travel agencies.

In contrast to the unfavorable Kayseri example mentioned earlier, Saratlı exemplifies effective protection and income generating within a limited budget.

The employment of rock-cut living clusters and cliff communities for commercial activities, including stores, restaurants, and hotels, has been suggested as a protective measure. This method, while neither culturally or ethically optimal, is seen as a final feasible alternative and has been executed in specific areas of Cappadocia and other regions globally.

However, it should be noted that the function of an underground shelter or a rock-cut church cannot be altered. The only possible functions for these structures are their own construction purposes.

In this context, revisiting our initial inquiry; if options are limited, how can we safeguard and conserve the over 300 rock-cut churches and more than 250 underground shelters for future generations?

The answer to this question is simple; unfortunately we can not!

There is almost no method to preserve and protect such a large number of rock-cut churches or underground shelters. As we wrote above, not only the inadequacy of the ministry budget or staff, but also many municipalities are indifferent and insensitive to such projects. From a commercial standpoint, the restoration and protection of all these structures is entirely ineffective.

Our experience in this field has painfully demonstrated that even a solid door for the protection of a cultural heritage is insufficient if there is no guard. In the course of our ongoing project in Kayseri, we constructed an iron door to safeguard the frescoed Isbidin Church from treasure hunters. However, this measure proved to be ineffective, as the door was breached a week later (Straub et al., 2019).

We observed that even opening them to tourism is not a solution unless they have a good location and intensive marketing. The Emiruşağı and Pazarören shelters in Kayseri were opened to tourism by local residents but have been neglected due to inadequate promotion.

Several scholars who have authored essays on this topic have underscored the necessity of safeguarding these structures by highlighting their cultural and historical significance. Some researchers have even written that these structures should be culturally preserved as soon as possible, based on the idea that underground cities are not only physical structures, but also representatives of social memory, identity and cultural diversity. On the other hand, the authors have failed to suggest any strategies for attaining this protection and preservation given the limited resources (Ayhan, 2004; Günay, 2024; Kahya Yıldırım, 2025; Kocalar, 2018) and many of these rock-cut churches and underground shelters have already suffered extensive damage over the past millennia, either from natural destruction or from treasure hunters.

We find it beneficial to maintain a realistic perspective regarding the restoration and protection of these structures. The solutions that can be proposed are restricted as a result of the aforementioned factors, including budget and personnel shortages.

Rock-cut dwellings can be repurposed for different functions. Nonetheless, this is impossible for the churches and subterranean shelters. Consequently, the only option in this situation is to immediately restore and preserve the most vital structures that remain unprotected. Unfortunately, the remaining hundreds of churches and subterranean shelters will be abandoned to their destiny.

References

Andaloro, M. (2010). *The project on the rock paintings in Cappadocia. The Church of the Forty Martyrs in Şahinefendi (Report 2008).* Paper presented at the 27. Araştırma Sonuçları Toplantısı, Denizli.

Andaloro, M., Valentini, V., & Pogliani, P. (2017). *Painting on rock settlements in Cappadocia. A database and a virtual museum for the enhancement of the area.* Paper presented at the HYPOGEA, Cappadocia (Turkey).

Ayhan, A. (2004). *Geological and morphological investigations of the underground cities of Cappadocia using GIS* (Ph.D.). ODTU, Ankara.

DÖSIM. (2025). *Döner Sermaye İşletmesi.* https://dosim.ktb.gov.tr/

Günay, G. (2024). Byzantine rock-cut architecture in Cappadocia and beyond: The state of scholarship. *Valonia, 1*, 35–63.

Kahya Yıldırım, E. (2025). Evaluation of underground cities in Nevşehir and its surroundings in terms of cultural geography. *Asia Minor Studies, 13*(1), 77–105.

Kocalar, A. (2018). Archaeological underground settlements and conservation problems: Nevşehir underground cities (Derinkuyu, Kaymaklı). *Journal of Awareness, 3*(3).

Straub, R., Yazlık, B., & Yamaç, A. (2019). Die Untergrund-Christen - Isbidin Kaya Kilisesi, eine stark gefahrdete Hohlenkirche in Kappadokien. *Mitteilungen des Verbandes der deutschen Höhlen und Karstforscher, 65*, 25–29.

Theophanis. (1839). *Chronographia* (Vol. 1). Bonnae (Bonn): Academiae Litterarum Regiae Borussicae.

Yamaç, A. (2019). Büyük Bürüngüz Underground Shelter (Kayseri–Turkey). *Opera Ipogea, 2–2019*, 65–76.

Yamaç, A. (2021). Rock-cut Byzantine churches of Koramaz Valley (Kayseri–Turkey). *Opera Ipogea, 1*, 51–68.

Glossary

Caldera Crater or surface depression resulting from the collapse of an underlying magma chamber roof during the withdrawal of magma, mostly related to ignimbritic eruptions for silicic magmas.

Cliff Settlement A cliff settlement means a rock-dwelled village with all the buildings, houses, storage rooms, barns, cisterns, even religious buildings, and defense shelters. The term "wall settlement" is used for structures dug on top of each other on a steep wall, and "cliff settlement" is used for settlements with more inclined slopes. In some different sources, all such structures are named "cliff dwelling" or "rock settlement."

Cone Dwellings A set of rooms intended for residential use dug inside the rock pinnacles that characterize various areas of Cappadocia. These rooms, even on different levels, are sometimes also used for other purposes, such as rock dovecotes, rock churches, and rock tombs.

Cones Erosion forms characteristic of Cappadocia. They represent the final product of the degradation processes of the pyroclastic slopes that have operated through thermoclastic erosion, cryoclastic erosion, the flow of the washing waters and the action of the winds. They have variable shapes, from conical to columnar, arranged individually or in groups along the valley floors or the valley slopes. Locally they are called "peri bacaları," that is, fairy chimneys, a term sometimes limited to indicating cones surmounted by a more coherent rock cap. Many of these natural structures have been excavated internally by man for use as rock dovecotes, rock tombs, rock churches, and conical dwellings.

Defense Devices Particular arrangements of an underground area aimed at protecting the occupants from enemy attacks. They could be active or passive (the latter being the most numerous). Passive defense devices are millstone doors, shield doors, joints, bottlenecks, pit traps, and angles. Active defense devices are peepholes, slits, blind tunnels, and escape routes.

Dovecote A structure intended to house pigeons or doves. They generally contain pigeonholes for the birds to nest. These structures facilitate the accumulation and collection of pigeon manure. The pigeon manure collected in this way has been used both in agriculture and in gunpowder throughout history.

Escape Tunnel Often, from the innermost rooms of an underground shelter, tunnels begin that are lower and narrower than those normally used, protected by smaller millstone doors. These could be used to evacuate the redoubt in the event of an enemy attack and, perhaps, to surprise the enemy from behind.

Fairy Chimney Literal translation of the Turkish term "peri bacaları." A term widely used locally to indicate the pinnacles or natural cones of tuff.

Fault Fracture of the earth's crust with movement of the two parts in contact along a vertical or inclined surface.

Ignimbrite Pyroclastic deposits primarily formed by volcanic ash and pumice, resulting from explosive eruptions that generate pyroclastic flows. Welded or unwelded, pumiceous, ash-rich deposit of pyroclastic density current. This term was formerly used for strongly welded deposits only. Ignimbrite is a pyroclastic flow composed of a very poorly sorted mixture of volcanic ash, pumice, and rock fragments. It is also called "tuff."

Isolating Room A room set up in the middle of an underground network that gives access to different areas or rooms that are separate from each other. Each room is protected by millstone doors that keep enemies out no matter what path they take, effectively cutting off whole areas.

Isolated Area Areas designed to ensure a community's temporary residence in the event of an enemy attack, communicating with similar areas via a central space (isolating room) that could be defended in opposing directions to isolate the various sectors based on the attackers' direction of penetration.

Joints Grooves dug into the walls of the tunnels intended for the installation of wooden barriers consisting of actual doors or of overlapping logs. This type of passive defense device is quite rare in Cappadocia due to the historical scarcity of wood in the region, which has led to the preference for stone closures called millstone doors.

Lava Liquid or highly viscous mass of magma, erupted by volcanoes with subsequent cooling outside the earth's crust.

A. Yamaç and E. Tok, *Underground Shelters of Cappadocia, Turkey*, Advances in Karst Science,
https://doi.org/10.1007/978-3-032-16095-9

Lava dome A lava that cannot flow due to high viscosity and accumulates over the vent.

Loopholes Holes made in the walls or ceiling of tunnels that lead to isolated areas and controlled by guard rooms, used to target enemies by blocking their advance. These are therefore active defense devices.

Magma A mantle- or crust-derived, physically and chemically complex mixture of molten rock, solids (e.g., crystalline or refractory fragments), and gases (e.g., CO or HO).

Millstone Door Defense devices characteristic of the underground structures of Cappadocia, in which, with rare exceptions, they are always present. Underground cities excavated for defense used circular stones, similar to millstones, to block tunnels and prevent enemy entry. In Cappadocia there are extreme samples with a diameter of 210 cm, a thickness of 60 cm, and a weight of 5000 kg. They are made up of large rock monoliths, shaped like a wheel or millstone. They were easily maneuverable in the appropriate maneuver chambers equipped with various types of doors. They generally have a central hole (peephole). They were used, in the event of an enemy attack, to block entire sections (isolated areas) of underground or rock settlements.

Narthex An architectural element typical of Byzantine churches consisting of the entrance or lobby area, located at the west end of the nave. The (pre-baptized Orthodox) and non-Orthodox are to stand here.

Naos The name was given to the whole middle and side naves of a church. Naos is the body of the church, consisting of a single aisle or a central aisle and side aisles.

Nave The main place in the architecture of the church, extending perpendicular to the apse, separated by columns or rows of pillars and where people pray. The nave is the main body of the church where the people stand during the services. This structure is divided into the middle nave and side naves.

Niche A functional cavity incised into a wall, possibly for placement or storage of objects, since most niches are at approximately chest height.

Open Courtyard Type of rock courtyard. In an "open courtyard," the central space is enclosed on three sides by rock walls in which the entrances to the underground rooms are located, while the fourth side is open to the outside (U-shaped courtyard).

Operation Room A small room excavated behind a millstone door, on the side of the tunnel, to close or open those millstone doors in underground cities.

Peepholes Holes drilled in the center of the millstone doors that could be used not only to observe the movements of strangers in the tunnels of the underground network but also to possibly hit them, thus becoming active defense devices. There was not even a need for lighting, considering that, due to the narrowness of the tunnel, the enemy had to proceed in single file and did not have the space to move out of the line of fire.

Pyroclasts Fragmentary material ejected during a volcanic eruption, including pumice, ash, and rock fragments.

Pyroclastic Deposit Deposit of volcanic origin produced by explosive eruptions consisting of a mixture of gases at high temperature with ash, fragments of lava, and pieces and blocks of rock called lithics.

Redoubt A "last defense or refuge room," the final chamber of an underground city.

Rock-Cut Church Underground structures intended for worship, built from a rock face, similar in layout to Byzantine churches built on the surface. These buildings appear to have been constructed "in negative" (i.e., by removing the rock) in all their architectural elements, including those with purely decorative value. Rock churches are found dug into tuff cones, in vertical rock walls, or associated with other rock structures.

Rock Courtyard Open-air space that serves as a common area for the surrounding underground dwellings. It can consist of a natural recess in a relief or be partially or totally excavated by man.

Rock-Cut (rock-hewn) General term for all human-excavated structures in a bedrock.

Rock "court" settlements Hypogeal settlements that develop around open-air rock courtyards, carved into the rock, whether they are located in a natural recess or specially dug by man. This type of space is distinguished depending on the morphology. In relation to the intended use of the underground buildings, we have the rock courtyard dwelling (or residence) and the rock courtyard monastery. The set of several courtyard buildings, combined or typology, gives rise to the rock courtyard village.

Rock Dwelling Name given to structures dug into bedrock and mostly used as a house.

Rock Settlements A category distinct from underground structures. Rock structures are those structures whose spaces dug by man into the rock, even on multiple levels, are all located, with a few marginal exceptions, above ground level, within elevations of the ground (vertical walls, cones, towers, pinnacles, etc.), and have a development limited to the part of the rock closest to the surface (first band). The road network that eventually connects the various units develops mainly outside the underground structures. Various "rock" typologies are recognized: cone-shaped dwellings, cone-shaped villages, wall villages, castle villages, "courtyard" villages, "courtyard" monasteries, rock churches, rock tombs, and rock dovecotes. Occasionally the structures indicated above are associated with underground shelters.

Rock Tombs In Cappadocia there are two fundamental types of rock tombs: pit tombs and chamber tombs. The former consists of rectangular pits large enough to accommodate a body, usually dug into the rock at the top of the buttes or into the floor of underground rooms attached to rock churches. The latter are chambers, sometimes with stone benches, also used for multiple burials, dug into vertical rock walls or erosion cones. May correspond to burials (tumulus) or archaeological hills (höyük).

Shaft The term is used here not with the meaning of "flue" but rather as an ascending vertical conduit, in the shape of a well, sized to allow the passage of a person, usually equipped with footrests, used to reach rooms dug at higher levels.

Shield Doors Much rarer than millstone doors and of more modest dimensions, they are rectangular stone slabs. They were perhaps used to prevent animals from leaving their underground shelters or as a marginal defensive element at the exit of small tunnels (passive defense devices).

Stratovolcano A volcano constructed of alternating layers of lava flows and pyroclastic rocks.

Tectonics Study of the structural characteristics of deformations and movements of the earth's crust with respect to stratigraphic and lithological aspects. The analysis of fold faults.

Troglodyte This word, which was used to mean a human cave dweller, comes from the ancient Greek *trogle,* "hole, mouse-hole," and *dyein,* "go in, dive in." Today, the word also signifies "living underground."

Tuff Generic term indicating a pyroclastic rock. In particular, it refers to a pyroclastic rock composed of lapilli (lava fragments) with a grain size ranging from 4 to 32 mm and subordinately by ash.

Tunnel Artificial tunnel with transverse dimensions rarely larger than those necessary and sufficient for the transit of an individual, almost always in a bent position. In secondary tunnels (escape routes), the passage is even more difficult and one must proceed on all fours. The underground structures of Tilkoy, Güzelöz, Tatlarin, and Kavlaktepe are distinguished by being equipped with a main tunnel of unusual dimensions both in width and height that we will call road tunnels.

Underground City A complex of underground structures located in many different parts of the world, used by locals to hide during an attack or raid. In the examples in Cappadocia, the tunnels of the buildings were closed with millstone doors and additionally protected by traps. These defense structures, whose small and singular ones are also named as "Underground Shelter" or "Underground Refuge," are thought to have merged over time to become "Underground City."

Underground Water Systems Structures excavated by man in the subsoil of canyons, especially in the area of erosion valleys, intended for the reclamation of stream beds and/or irrigation of cultivated terraces. The most important works are represented by the underground drainage channels/ducts, long tunnels dug along the side of the valleys, following their trend and slope. Their function is to capture and transport torrential water, avoiding erosion of the riverbed, which can then be usefully cultivated. The underground collection channels/ducts are more modest tunnels in length and section. They are usually dug perpendicularly to the walls of the canyon until they reach the aquifer, which is then collected in larger spaces, the underground cisterns, from which the water can be drained to be used for irrigation purposes. In this way they perform the function of wells, which, in the area of the erosion valleys, are poorly documented.

Underground Settlements See underground structures. Note that the term "underground" is often used in a generic sense and, in this case, also includes rock structures.

Underground Structures Distinct from rock structures. Note that the term "underground" is often used in a generic sense and sometimes also includes rock structures. By underground structures, however, we mean, in a strict sense, those excavated by man in the innermost areas of rocky reliefs (inner band) or in depth, below the ground level, in vertical bands (epidermal band, intermediate band, deep band). The road network connecting the various underground units is mostly found underground, thus constituting complex. We recognize various "underground" typologies, such as underground cities, underground shelters, and underground water systems.

Underground Shelters A particular case of underground structures that generally extend over a single horizontal level in the innermost band of the short vertical slopes that surround the buttes. They have a predominantly defensive use. They are usually equipped with a redoubt, escape routes, and multiple defensive devices (grindstone doors, shield doors, blind tunnels, pit traps, narrow passages, slits, etc.). Although autonomous in terms of water supply and, presumably, the abundant food stored in the silos, it is believed that they were inhabited for periods of time that were not too long and/or in the event of danger from bands of raiders. They are sometimes found combined with other rock structures.

Wall Settlement Structures, mainly for residential purposes, were dug into the overhanging walls of vertical slopes. They are composed of independent units, made up of several underground rooms, often located on overlapping floors that rise from the ground level. The various dwellings that make up the "wall village" are usually connected by an external road network whose main routes are identified with the axes of the valleys themselves.

Index

A
Abbasid Caliph al-Mahdi, 133
Abuşağı, 50
Acigöl, 7, 47, 48, 50, 51, 55, 56, 80, 87, 109, 127
Acıpınar, 47
Ağadam, 47
Ağcaşar, 50
Ağıllı, 33, 47, 56
Ağırnas, 35, 47, 50, 51, 57–59, 65, 106
Ahlat, 16, 17
Ahurian, 16
aisle, 85, 121, 138
Akçakent, 49
Akdam, 47
Akdere, 69
Akhisar, 49
Akköy, 5
Akmezar, 47
Aksaklı, 47
Aksaray, 2, 10, 13, 22, 30, 47–51, 66, 89, 120, 134
Aktaş, 47
Aktepe, 47
Alakuşak, 70
Alanyurt, 47
Alayhanı, 47
altar, 85
Altunhisar, 50, 51
Altınkaya, 49–51
Altınyazı, 47
Anatolia, 1–3, 5, 8–11, 13–17, 19–21, 31, 32, 34, 36, 63, 95, 133
Andesite, 8
Ani, 14, 16, 17, 31
Ankara, 34, 36
Antiochus, 11
apse, 18, 83, 85, 94, 120, 138
Arab, 2, 4, 13, 15, 17, 18, 31, 32, 36, 95, 133
Archelaos, 12
arcosolium, 127
Ardahan, 16
Argaeus, 7, 12
Ariarathes, 11, 12
Ariobarzanes, 12
Armenia, 12, 13
Assyrian, 11, 34
Aşıklı, 10
Augustus, 12
Avanos, 22, 47–51, 64, 114
Avladağ, 10
Ayan, 47
Aydıntepe, 35, 36
Ayşepınar, 17, 18
Ayvalı, 83
Azıklı, 48

B
Babakonağı, 49
Bağlıca, 48
Bağlıkaya, 48
Bahadın, 48
Balak, 50
barrel-vaulted, 57, 94, 106, 107, 120, 123
basalt, 66
Basil, 13, 15
basilica, 85, 96
Batman, 15
Bayburt, 36
Bebek, 48
Bektaşlı, 50
Belağası, 48, 51, 61–63, 114
Belha, 48, 64, 65, 114
Belisırma, 48, 50, 129
Bezirhane, 20, 48
Bitlis, 15
Boğazköy, 48
Boğazlıyan, 47, 50, 51, 60, 61
Bölükören, 49
Bozcatepe, 49
Bozcayurt, 49
Bozköy, 51
Bünyan, 49, 50
Bürüngüz, 1, 29, 31, 36, 37, 39, 49, 65–67, 134
Büyükgüve, 49
Byzantine, 3, 5, 13–19, 21, 22, 31, 32, 35, 36, 61, 71, 83, 96, 101, 102, 133, 138

C
Caesarea, 12–15, 32
Çağıl, 49
Çakmaktepe, 10
Çakıllı, 49, 73
Çakıltepe, 10
Caldera, 6–9, 137
Çamardı, 50
Camiliören, 49
Canavar, 49
Çankırı, 36
Çanlı, 49

A. Yamaç and E. Tok, *Underground Shelters of Cappadocia, Turkey*, Advances in Karst Science,
https://doi.org/10.1007/978-3-032-16095-9

Cappadocia, 1–5, 7, 8, 10–22, 25–37, 39, 40, 44, 45, 47, 48, 55, 57, 65, 66, 85, 88, 94, 103, 105, 106, 109, 111, 120, 121, 128–130, 133, 134, 137–139
Çardak, 7, 49
Castle village, 17, 27, 139
Çatalhöyük, 7, 8
Çatalin, 49
Çatalsu, 49
Çavuşin, 2, 30, 51
Çayırlı, 49
Çekiçler, 49
Çeltek, 49
Cemil, 49, 50, 133
Cemilkoy, 7
Çerkezbayın, 49
Çeşka, 47–49
Cevizli, 89
Cevriye, 50
Chalcolithic, 10
chimney-vault, 58
Christianity, 5, 17
Church, 2, 4, 5, 13, 15–22, 27, 29–31, 35, 37, 48, 50, 56, 57, 59–61, 71, 76, 82, 83, 85, 87, 92–96, 118, 120, 121, 123, 127, 133–135, 137–139
Cicero, 12
Çiftlik, 51
Çiftlikköy, 49
Cilicia, 12
Cistern, 4, 15, 27, 133, 137, 139
cistern, 59, 60, 62, 74, 84–86
Çolaknebi, 49
Çomaklı, 40, 49, 67, 69
Çömlekçi, 49, 69, 70
Cone-shaped, 30, 139
Çorum, 15
Croesus, 19
Çukurören, 49
Cyrus, 19
Çıralı, 30, 47–49, 66–68

D
Dacitic, 6, 8
Damlalı, 49
Danalık, 35, 49, 69, 70
Darius, 11
Değirmen, 49
Demirci, 49
Derinkuyu, 1, 7, 25, 26, 28, 29, 31, 36, 47, 49–51, 63, 65, 70–73, 75, 76, 86, 93, 95, 109, 112, 115, 126, 134
Develi, 5, 8, 10, 49–51, 67, 124
Dikilitaş, 49
Dikkartin, 9
Dikmen, 51
Doğala, 49, 73
Doğanlı, 35, 49, 74, 75
Doğantarla, 49
Dolichenus, 36
Dorukini, 49
Dovecote, 4, 16, 17, 19, 27, 29–31, 60, 105, 109, 124, 128, 129, 137, 139
Düğen, 51
Düğüz, 49
Dülük, 15, 36
Durmuş, 49, 75–77

E
Edessa, 12
Edikli, 49
Elmacık, 49
Emiruşağı, 49, 76, 78, 134
Erciyes, 5–10, 20, 96
Erdas, 73
Ereğli, 11, 12
Ergani, 36
Erkilet, 6, 8, 10
Eskil, 48, 49
Eskinuz, 51
Eskişehir, 9, 15, 96
Eşmekaya, 49
Eumenes, 11
Euphrates, 12, 16, 31
Eusebeia, 12
Eusebes, 11, 12
Eustachius, 51

F
façade, 15, 90, 97, 98
Fakılı, 49
Fatinler, 49, 79–81
Fault, 5, 9, 10, 139
Filikören, 28, 49, 80
Filiktepe, 29, 31, 35, 39, 49, 80, 82, 83, 118
Fresco, 13, 19, 21, 133
fresco, 95
Fırkatan, 49
Fırıntaş, 49

G
Gaziantep, 15–17, 36, 43
Gaziemir, 49
Gelegüle, 49
Gelesin, 49
Gelveri, 49, 89
Genezin, 51, 114
Germanicus, 12
Germir, 49
Gesi, 10, 48, 49, 51, 61
Gökçe, 49, 50
Gökçetoprak, 51, 118
Göktaş, 49
Gölcükbaşı, 49
Golgoli, 30, 35, 40, 49, 83–86
Golgotha, 84
Göllü, 5
Göllüdağ, 7
Gomeda, 49, 85–87
Gordian III, 12
Gordion, 34
Göre, 49
Göreme, 2, 13, 17, 19, 21, 30, 46, 48–51, 58, 85, 86, 103, 128, 129
Göstesin, 25, 29, 31, 49, 87–89, 118
Göynük, 49
Gözkayası, 49
Gözlükuyu, 49
Gülağaç, 47, 49, 51
Gülbayır, 49
Gülistan, 49
Gülpınar, 47, 49

Gülşehir, 29, 47, 49–51, 91, 93, 109
Gümüşkent, 49
Gürlek, 49
Gürpınar, 125
Gürün, 11
Güzelköy, 51, 125
Güzelöz, 30, 49, 88, 90, 91, 139
Güzelyurt, 47–51, 89, 91, 92, 109

H
Hacıbektaş, 33
Hacıgaybı, 49, 51
Hadrian, 12
Halfeti, 16, 17
Halys, 11, 19
Harman, 49
Hasandağ, 7, 8
Hasankeyf, 16, 17
Hasanköy, 49
Hasanlar, 10
Hasinli, 47, 60, 61
Hattusa, 3, 36
Herodotus, 3, 19
Hilar-Çayönü, 36
Hittite, 3, 11, 32, 33, 56, 61, 73
Höbek, 50
Hromgla, 16
Hıcıp, 47, 49

I
İğdeli, 47
ignimbrite, 5–8, 84, 118, 137
Ihlara, 6, 22, 30, 50
Ilısu, 50
İncesu, 6, 8, 50
İncili, 50
İnözü, 50

J
Justinian, 15

K
Kadı, 50
Kaletepe, 10
Kalfat, 36
Kangal, 11
Kanis, 10, 11
Karaağaç, 50
Karaasmaz, 50
Karabulut, 30, 85, 128
Karaburna, 33, 34, 50
Karacalı, 50
Karagüllü, 9
Karakaya, 50
Karakoç, 50
Karaman, 16
Karanlık, 50
Karapınar, 30, 49, 50, 66
Karataş, 50, 51
Kargın, 50
Kars, 16, 31
Kasımlı, 50
Kavak, 7
Kavlaktepe, 35, 50, 139
Kayabağ, 49, 51
Kayabaşı, 50
Kayakapı, 37
Kaymaklı, 25, 26, 29, 31, 37, 50, 63, 65, 93–97, 103, 115, 134
Kayseri, 2, 5, 7, 9–11, 13, 14, 17, 18, 20, 22, 26, 27, 30, 31, 35, 39, 40, 47–51, 57, 59, 61, 65–67, 76, 79, 88, 96, 98, 101, 106, 116, 118, 124, 125, 133, 134
Kayırlı, 50
Kaz, 50, 73
Keçiboyunduran, 5
Kemer, 50, 95, 98
Kemerhisar, 11
Kemerkaya, 86
Kepez, 49, 50
Keşlik, 50, 133
Kilistra, 15
Koç, 8
Kocasinan, 50
Koçpınar, 50
Koloneia, 22
Komana, 12, 13
Kömür, 50, 99, 101, 103, 104
Kömürcü, 10
Konaklı, 14, 50
Konya, 7, 15, 30, 47, 49–51, 66, 68
Koramaz, 17, 57, 65
Köseyusuflu, 50
Köşk, 10
Köyiçi, 36
Ktistes, 12
Küçükgüve, 50
Kulluk, 50
Kulpak, 50
Kültepe, 10, 11
Kümbetaltı, 50
Kumluk, 51
Kumtepe, 7
Kura, 16, 17
Kurugöl, 51, 55, 109
Kuruköprü, 50
Kuşcağız, 50, 101, 102, 105, 106
Kuyulutatlar, 50, 109
Kybistra, 11, 12
Kılıçlar, 21, 50
Kırağcaşar, 50
Kırkgöz, 50, 123, 134
Kırlangıç, 30, 50, 96, 98–100
Kırşehir, 2, 13, 30, 47, 48, 50, 108
Kışla, 10
Kızıl, 8
Kızılkaya, 7, 50
Kızılören, 50
Kızılırmak, 6, 11, 19, 109, 118

L
Lava, 5–9, 138, 139
Lefkere, 49
Leo, 31
Loophole, 38, 39, 42, 138
Luwian, 33, 34
Lykaonia, 12
Lysimachus, 11

M

Mahmatlar, 50
Mahmutçuk, 50
Mamasun, 50
Manastır, 48–50
Manazan, 16
Mandama, 49
Mantzikert, 14, 32
Mataza, 103
Mazaka, 12
Mazı, 50, 86, 103, 107
Mecemir, 50
Melendiz, 5, 50, 129
Meskendir, 30, 50, 86, 103, 105, 106, 108, 129
Miocene, 5, 7
Mucur, 30, 47, 48, 50, 108, 111
Munamak, 49
Mustafapasa, 49, 83, 85

N

naos, 57, 85, 138
narthex, 57, 85, 138
Nave, 138
Neogene, 5, 7, 66
Neolithic, 7, 10
Nevşehir, 2, 5, 10, 11, 13, 14, 20, 22, 25, 26, 33, 47–51, 55, 70, 80, 83, 87, 93–95, 115, 134
Niğde, 2, 10, 11, 13, 14, 22, 47, 49–51, 94
Nikephoros, 31

O

Ören, 25, 50, 51, 95, 98
Örentepe, 25, 26, 51, 109, 111, 112
Ortahisar, 17
Ortaköy, 50
Osiana, 96
Ötedere, 37, 51, 62–64, 112–114
Ovakent, 60
Ovaören, 25, 40, 47, 49, 50, 80, 82, 87, 89, 95, 118
Oymalı, 31, 35, 50
Ozancık, 50
Özkonak, 26, 48, 51, 64, 114–116
Özler, 51
Özlüce, 51, 115, 117

P

Parzuta, 33
Pazarören, 47, 48, 51, 116–118, 134
Penzikli, 51
Persia, 11, 43
Philadelphos, 12
Phrygian, 15–17, 32, 34
Pontus, 11, 12
Pörnek, 51
Pusatlı, 49, 51, 77, 79, 80
Pyroclastic, 5, 7, 16, 137–139
Pınarbaşı, 51, 116

Q

Quaternary, 5–7

R

refectory, 58, 59, 71
Roman, 3, 5, 7, 11–13, 15, 17, 22, 31, 36, 101–103

S

Sağırkaraca, 51
Şahinefendi, 19, 50, 51, 133
Şahmelik, 51, 124, 125
Saratlı, 50, 51, 120, 134
Saraycık, 51
Sarıağıl, 51
Sasanian, 15, 32
Sebasteia, 14, 32
Seleucid, 11
Selime, 13, 49, 51, 91
Seljuk, 13, 14, 21, 32, 61, 73, 93, 120
Sesverenpınar, 36
Sevinçli, 49
Şeyhler, 51
Sille, 15
Sinan, 106, 107
Sinassos, 83
Sivasa, 26, 28, 29, 33, 34, 36, 39, 51, 118, 119, 121
Sobesos, 11
Sofular, 7, 10, 51
Soğanlı, 6, 21, 49
Solaklı, 108
Sorgun, 48
Sosun, 51
Strabo, 4, 9, 19
Stratovolcano, 7, 139
Subaşı, 51
Susadı, 51
Şıh, 49, 79

T

Talas, 47, 49–51, 69
Talasıra, 51, 125–127
Taniktepe, 118
Taşpınar, 51
Tatlarin, 29, 30, 50, 51, 91, 127–129, 139
Tatlıca, 51
Taurides, 5
Tectonic, 9, 10, 139
Tekkaya, 30, 51
Tepebağları, 10
Tilköy, 51
Tokarız, 51
Tomarza, 49–51, 76, 77, 79, 98, 99, 101
Tombak, 51
Topada, 33, 34, 47, 56, 73
Topakhöyük, 10
Topakkaya, 8
Topuzdağ, 7
Troglodyte, 2, 4, 5, 16, 17, 19, 27, 31, 139
Tuwana, 11

U

Uçhisar, 17, 19–21, 44
Ürgüp, 5, 13, 19, 20, 22, 30, 37, 38, 41, 45, 49–51, 83
Uzunkaya, 51

V
Valens, 13, 15
Valerian, 12
Vespasian, 12
Viranşehir, 49

W
Warpalawa, 11
Wasusarma, 33

X
Xenophon, 31

Y
Yağan, 51
Yallı, 51, 64, 65, 114
Yarlağan, 51
Yassıhöyük, 10
Yenice, 50
Yenifakılı, 49
Yeniköy, 17
Yenipınar, 49, 51
Yeniyuva, 51
Yeşilgölcük, 49
Yeşilköy, 50
Yeşilöz, 49, 51
Yeşilyurt, 51
Yozgat, 2, 13, 47–51, 60
Yunak, 51
Yılanlı, 9
Yıldıztepe, 51

Z
Zeila, 115
Zelve, 2, 7, 51, 134
Zemi, 30, 51, 85, 128, 130
Zile, 51, 115, 117
Zincidere, 51